Interdisciplinary Approaches to Environmental Identity and Portrayal

Interdisciplinary Approaches to Environmental Identity and Portrayal

Edited by

Mohamed Mliless, Mohammed Moubtassime,
and Diego L. Forte

BLOOMSBURY ACADEMIC
NEW YORK • LONDON • OXFORD • NEW DELHI • SYDNEY

BLOOMSBURY ACADEMIC
Bloomsbury Publishing Inc, 1359 Broadway, 12th Floor, New York, NY 10018, USA
Bloomsbury Publishing Plc, 50 Bedford Square, London, WC1B 3DP, UK
Bloomsbury Publishing Ireland, 29 Earlsfort Terrace, Dublin 2, D02 AY28, Ireland

BLOOMSBURY, BLOOMSBURY ACADEMIC and the Diana logo are trademarks of Bloomsbury Publishing Plc

First published in the United States of America 2025

Library of Congress Cataloging-in-Publication Data Available

ISBN: HB: 978-1-66698-021-9
PB: 979-8-21637-678-1
ePDF: 979-8-76516-512-6
eBook: 979-8-76515-988-0

Typeset by Deanta Global Publishing Services, Chennai, India

For product safety related questions contact productsafety@bloomsbury.com.

To find out more about our authors and books visit www.bloomsbury.com
`and sign up for our newsletters.

Contents

Figures

Tables

Foreword

William R. Naugle

When I first traveled to Morocco, I was uncertain of what I would find there or how I would react to those whom I would meet and the sites that I would see. I fell in love. The culture transcends time: while you are in the middle of a modern, thriving society, at the same time, you are thrown back several millennia through the rhythm of ancient cities and souks, a confluence of linguistic heritage, and the people who seem to hold the knowledge of civilizations past and present. But, culture, language, and people are somehow influenced and shaped by the geography, flora, and fauna, among which they live. Morocco proves breathtaking from these latter three aspects: from the Atlas Mountains, to the stormy Atlantic pounding upon the shores of Rabat, to the sands of the Sahara, to the Maâmora forest, and to the warm waters of the Mediterranean coast. It is a sensual country that stirs the soul and convinces the traveler that there is something greater than all of these component parts, than any individual, or groups of individuals. And, it is humbling to learn, through this interaction of things great and small, that each of us not only forms a part of this miraculous ensemble, but also that we each have a responsibility to the environment, both societal and ecological, in which we live.

In *Language Society and the Environment: How the Environment is Shaped through Discourse*, Mliless and his colleagues have generated a volume that collates every aspect of the geo-socio-political conversation on the environment, with each section and chapter carefully laying out the human impact we have had and road-mapping the necessary steps to be taken to mitigate, if not reverse, our current course, through literary conversation, discussion, and debate. The text is a travelogue of intricate and finely woven linguistic, literary, and cultural fabric on full display, leading the reader to engage intellectually and scientifically with ecolinguistics, ecocriticism, and ecofeminism. Mliless weaves together the chapters masterfully, drawing the reader subtly through challenging concepts and difficult realities. In skillful placement of the final chapter, "*The Rain Sound as an Affective Force that Unites Humans and Nature*," he provides the reader with comfort and hope for a future harmonious resolution of the existential threat posed to ecosystems by the human element—the sound of rain draws us deep within our primordial selves to find their peace and, perhaps, solidarity.

Thus, the psychology of this compendium entreats faculty to assign it as part of the required reading necessary, at both the undergraduate and graduate levels across a spectrum of disciplines, to introduce students to the ongoing dialogue of the interaction and intersection of humanity and the environment. It will be useful for those in the hard sciences, as well as the humanities, helping students to understand their role in civil responsibility and as actors on and in nature. The genius of the book is that it is not simply an academic silo; rather, it is accessible to anyone interested in being introduced to or learning more about our multifarious relationships to the environment. While delving into water stress in Morocco, animal welfare in Argentina, military green defense in the United States of America, or the forest policy of India, the reader is not only a student of environmental impact, but an intrepid voyager experiencing an illusory, but indelible, sensory exhilaration of place.

The place that takes center stage is Morocco, where the smell of orange blossoms intoxicates, where mountains meet the sea, where camels have crisscrossed the Saharan sands on countless journeys, and where the people have lived on and struggled with the land for countless millennia. It is unsurprising, then, that Morocco should take great interest in playing the prospective role of arbiter in this discourse: Al-Maghrib (المغرب) is the west, where the sun sets and where the stars melt into the ocean—the quiet place of contemplation and reflection, where Fatima al-Fihri gave birth to the University of Al Quaraouiyine. Appropriately, Mliless and his collaborators guide us on a journey of intellectual curiosity, a journey that is sometimes gripping and sometimes soothing, but never disappointing. "*Language Society and the Environment: How the Environment is Shaped through Discourse*" stimulates our thoughts, stirs our imagination, and challenges us to embrace the responsibility humanity has to nature, so we can continue to discover the flora and fauna around us, so we can continue to fall in love with this wonderful creation of which we are part.

Acknowledgments

This work would not have been possible without the joint efforts of the prominent editors, namely Dr. Mohamed Mliless, Dr. Mohammed Moubtassime, and Dr. Diego L. Forte. We are especially indebted to Professor Dr. Mohamed Behnassi, a Moroccan Senior Environmentalist Expert at the Economic Social and Environmental Council (ESEC), for being there for us throughout the editing stage. We recall, in this context, the efforts of every participant whose diligence has contributed a lot to the achievement of this humble work in due time. I would especially like to thank Professor Dr. William R. Naugle, who wrote the Foreword of this book. I want to express my sincere gratitude for writing Foreword to our book.

Abbreviations

ACEC	Association of Cultural Ecology and Communication
ADEME	The French Agency for Ecological Transition
ADFM	Association Démocratique des Femmes du Maroc
AFED	Arab Forum for Environment and Development
AID	Agency for Innovation
AMDF	Association Marocaine des Droits des Femmes
APA	American Psychological Association
BFI	British Film Institute
CA	Content Analysis
CAN	Center for Naval Analysis
CBF	Central Board of Forestry
CDA	Critical Discourse Analysis
CELDF	Community Environment Legal Defense Fund
CERES	Center for Environment Human Security and Governance
CNAS	Centre for a New American Security
CODES	Coalition for Digital Environmental Sustainability
COP	Conference of the Parties
CSD	Campaign for Survival and Dignity
CSMD	Commission Spéciale sur le Modèle de Développement
CVI	Content Validity Index
DDT	dichlorodiphenyltrichloroethane
DEP	Drought Emergency Plan
DFD	Dark Firm and Dry
DHH	Deaf and Hard of Hearing
DoD	Department of Defense
ECDA	Environmental Critical Discourse Analysis
EDA	Ecocritical Discourse Analysis
EE	Environmental Education
EIC	East India Company
ESTK	Higher School of Technology in Khenifra
ESEC	Economic, Social and Environmental Council
EU	European Union
FRA	Forest Rights Act
GDP	Gross Domestic Product
GEEP	Global Environmental Education Partnership

GEMS Air	Global Environment Monitoring System for Air
GHG	Greenhouse Gas
HCP	Haut Commissariat de Planification
HTU	Hussein Technical University
IASCAV	Instituto Argentino de Sanidad y Calidad Vegetal
IEA	International Ecolinguistics Association
IPCC	Intergovernmental Panel on Climate Change
ITT	Ishpingo-Tambococha-Tiputini
JFM	Joint Forest Management
LDDF	Ligue Démocratique des Drois des Femmes
MAD	Moroccan Dirhams
MENA	Middle East and North Africa
MFP	Minor Forest Product
NAFAC	Naval Facilities Engineering Command
NATO	North Atlantic Treaty organization
NBI	Nature-based health interventions
NCA	National Commission on Agriculture
NDCs	Nationally Determined Contributions
NDP	New Development Plan
NEEF	Environmental Education Foundation
NSF	National Science Foundation
OSCE	Organization for Security and Cooperation in Europe
PNUE	Programme des Nations unies pour l'environnement
SENASA	Servicio Nacional de Sanidad y Calidad Agroalimentaria
ST	Source Text
UAF	Union de l'Action Féminine
UN	United Nations
UNFCCC	United Nations Framework Convention on Climate Change
UNICEF	United Nations Children's Fund
UNEP	United Nations Environment Programme
UNESCO	United Nations Educational, Scientific and Cultural Organization
VEC	Visual Environmental Communication
WECF	International: Women Engage for a Common Future
WGII	Working Group II

Introduction

Mohamed Mliless

Interdisciplinary Approaches to Environmental Identity and Portrayal is a comprehensive exploration of the multifaceted relationship between humans and their environment. This book delves into the intricate web of connections, influences, and reflections that shape our understanding and interaction with the world around us.

The book is a scholarly exploration that crosses multiple disciplines to provide a comprehensive understanding of how we perceive and portray our environment. It begins with an in-depth analysis of the narratives that shape our understanding of food production and animal welfare, and their implications on our relationship with the environment. This is followed by a discussion on the unconventional role of the military in mitigating climate change, offering a unique perspective on environmental stewardship.

The narrative then delves into the linguistic dimensions of our interaction with the environment, examining how language reflects our perception of landscapes as sentient entities. It further explores the convergence of environmental protection and the digital realm, underscoring the pivotal role of technology in environmental conservation.

The book also meticulously covers the legal aspects of environmental protection, including the laws governing hazardous waste, sustainable development, and water rights. It provides an intricate case study of Morocco, discussing the balance between the right to a healthy environment and the interplay between international conventions and national legislation.

Historical perspectives are also provided, with a focus on forest policy and management in India during colonial and post-colonial times. The book also discusses the portrayal of hunted animals on social networks, providing a visual and critical discourse analysis.

The latter part of the book delves into the psychological and societal aspects of environmental interaction, discussing the role of the environment in shaping identity and society. It also explores the intersection of the environmental crisis, postcolonialism, and gender inequality in the Global South.

The book also explores the role of women in environmental conservation, providing a feminist perspective on environmental issues. It delves into the portrayal of environmental issues in literature, providing a critical analysis, and discusses the concept of ecotherapy and communion with nature, exploring the therapeutic benefits of nature and how communion with nature can contribute to mental health and well-being.

Each chapter of the book provides a unique perspective on environmental issues, making it a comprehensive resource for anyone interested in environmental studies. The book is a must-read for anyone interested in understanding the complex interplay between humans and their environment, from various disciplinary perspectives.

The Book's Thematic Approach

Chapter 1, "Global Climate Change: From Anthropocene to Capitalocene," authored by El Houcine Chougrani and Mohammed Adli, explores the link between climate change and capitalism in a world where environmental protection and resource conservation are vital for survival. The authors posit that future generations, like their predecessors, have the right to live in harmony with nature and establish positive connections, as this aligns with the principles of sustainability. Human activity has exerted a dominant influence on the climate and environment during the Anthropocene, a geological period that commenced with the "First Industrial Revolution" in 1750. However, it's worth noting that the third world, particularly the African continent, has contributed minimally to this phenomenon. While humans have traditionally valued harmony with nature and the balance of natural resources, capitalists have often overlooked nature's protection in pursuit of profit maximization. Thus, it can be argued that the Capitalocene, characterized by capitalist dominance, prevails in the economic sphere. From this perspective, nature has always been a tool or instrument (as suggested by Descartes) utilized by people to satisfy their sustained needs and achieve their objectives. The study focuses on the evolving relationship between humans and nature, tracing the journey from a time when humans were deeply connected to nature to the present, where this harmonious relationship appears to have been lost due to human selfishness and the desire to profit from natural resources. The research will shed light on the manipulation of nature, which will shape the future relationship between humans and nature. To facilitate this analysis, the research will also focus on the shift from the Anthropocene (encompassing all humans, all ages, etc.) to the Capitalocene (characterized by capitalist egoism). The aim is to demonstrate that all nations and individuals bear the responsibility for nature's protection, despite some people's belief that the Capitalocene lies at the heart of the escalating climate change crisis and will continue to bear its responsibility in this disaster.

Chapter 2, "Emotions and Disease: The Negative Framing of Non-Human Animals in Climate Change Discourses," authored by Diego Luis Forte, delves into the intersection of government policies, human interactions with nature, and the narratives that shape our relationship with the environment. The author posits that as citizens of a modern nation, our interactions with nature are regulated in various ways by government policies. These range from access to natural resources, the food we consume, the clothes we wear, to our perception of non-humans and nature in general as resources for our survival. Forte explains that official agencies, which control and regulate food production, blend scientific and commercial discourses. They transform their narrative into the official, legal story, validating it not just by social convention but also by law enforcement. The chapter aims to analyze the animal welfare manual currently in use by the SENASA (Servicio Nacional de Sanidad y Calidad Agroalimentaria), an Argentine agency under the Ministry of Agriculture, for cattle breeding fields and slaughterhouses operating at the national level. As a theoretical framework, the chapter adopts multimodal critical discourse analysis and uses the analytical tools proposed by Hodge & Kress (1993), Lakoff (1987, 2002), and Kress & van Leeuwen (2005). The chapter is part of a larger research project that analyzes how discursive patterns impact the professional practices carried out by the agents of SENASA. The working hypothesis of this study is that the combination of scientific and neoclassical economic discourses structures the semiotic resources in the manual to avoid explicitly mentioning human agency in processes that negatively impact non-humans and the environment.

Chapter 3, "Grammatical Reflections of Sentient Landscapes: Linguistic Evidence for Environmental Personhood," authored by David Stringer, delves into the linguistic aspects of our interaction with the environment, particularly the concept of environmental personhood.

Environmental personhood has emerged as a powerful legal concept in recent years, originally based in part on the definition of corporations as entities possessing juridical rights. However, while corporate personhood has for centuries served as a legal metaphor, environmental personhood is in fact literal for many indigenous communities: personhood is not restricted to humans, nor animacy to animals. In major environmental legal cases invoking personhood for enspirited landscapes on the basis of Indigenous beliefs, one set of arguments yet to be systematically advanced involves the encoding of features such as personhood in language. Environmental personhood may be linguistically encoded through pronouns, direct object marking, numerical classifiers, verbs of existence, or word-order hierarchies, In this paper, I contrast metaphorical and literal interpretations of legal personhood and suggest that grammatical reflections of animacy represent untapped potential for legal arguments advanced by communities seeking rights for sentient landscapes.

Chapter 4, "Preservation of the Environment and Development of Digital Technology," authored by Hind Majdoubi, explores the intersection of environmental protection and digital technology, highlighting the role of technology in environmental conservation. The author posits that the preservation of the environment and the development of digital technology are two crucial issues that often seem to be in conflict. On the one hand, we need to protect our planet and reduce our environmental impact to ensure its habitability for future generations. On the other hand, digital technology has the potential to revolutionize our lifestyle and work, and can aid us in addressing some of humanity's most significant challenges. One way to reconcile these two issues is to focus on developing environmentally sustainable digital technologies. For instance, we can invest in renewable energy sources like solar and wind power to supply the electricity needed for our digital devices. We can also design technologies that are more energy-efficient and produce less waste. Another approach is to use digital technology to help us monitor and protect the environment. For example, satellite imagery can be used to track deforestation, pollution levels, and the health of ecosystems. This data can then inform policies and interventions aimed at reducing environmental damage. Ultimately, the key to reconciling environmental preservation and digital technology development is to ensure that our technological innovations are guided by a strong ethical framework that prioritizes sustainability and environmental responsibility. By doing so, we can leverage the power of technology to create a more sustainable future for all.

Chapter 5, "The Law of Hazardous Waste and Sustainable Development," authored by Khadija Anouar, discusses the relatively recent issue of waste management in Morocco. It wasn't until the 2000s that the first controlled landfill was installed, and the first law governing the sector was issued. As Morocco rapidly expands, waste management is becoming an increasingly important issue in promoting a more sustainable environment. Like many other countries in the Global South, Morocco has committed itself to various regulatory and policy frameworks to meet the requirements of efficient and rational waste management. This commitment is centered on sustained efforts to adapt, strengthen, and renew the legal system that provides a framework for public action in waste management. However, much still needs to be done to recover waste, as only 10 percent of it is currently recovered for recycling. The chapter acknowledges that safe and accurate management of waste can only be implemented if reforms and norms are promulgated. The author notes that the law of hazardous waste has undergone a clear mutation since 2008, with new objectives pursued and a wide variety of situations now covered by hazardous waste treatment. The author hopes that this complexity will not complicate the law but simplify it for the benefit of all actors.

Chapter 6, "The Protection of Water Rights in the Face of Human Security: The Case of Morocco," authored by Rabiaa Bourhim, delves into the governance of natural resources, specifically water, in the context of increased vulnerability. The author posits that the governance of natural resources exploited in common by individuals presents a significant challenge and concern for both the state and local actors. In response to this, Morocco has implemented various public policies regarding the governance of common resources, particularly water. However, these policies face substantial problems and resistance from concerned actors due to questions about the legitimacy of the actions taken and the methodology adopted. The chapter describes two traditional processes, namely the collective management of water (Lkhettara, Tighira), that have demonstrated their resilience, legitimization, mutual control, and democratization in the management of common natural resources. Contrary to the proposals of the "tragedy of the commons," problems of implementation are sometimes resolved by voluntary organizations rather than by a coercive state. This chapter underscores the importance of collective management and voluntary organizations in addressing the challenges of water governance and protecting water rights in the face of human security.

Chapter 7, "The Right to a Healthy Environment in Morocco between International Conventions and National Legislation," co-authored by Fatima Zohra Alaoui Mahrez and Hind Hassine, discusses the balance between international conventions and national legislation in the context of environmental protection in Morocco. The authors highlight that the protection of the environment has become a crucial issue in all legal systems due to its significance across all sectors. Consequently, the Moroccan government has incorporated environmental protection into its public policies. Morocco has ratified numerous international conventions that guarantee the right to a healthy environment. Furthermore, the country's Constitution of 2011 underscores the critical role of the environment and establishes various laws for its preservation. Morocco has initiated several programs, projects, and institutions aimed at implementing these laws. The primary objective of this chapter is to spotlight Morocco's role in preserving the right to a healthy environment. The authors document, describe, and analyze the Moroccan legal experience in environmental protection. The main findings of this study are: firstly, Morocco has been an active participant in environmental protection at local, regional, and global levels; secondly, Morocco has demonstrated its commitment through the establishment of the National Council for the Environment and other institutions and programs; and thirdly, the Moroccan legislator should consolidate all laws related to the environment into one law under the name of the "Environmental Code" for easier research and reference. This chapter provides a detailed case study of Morocco, offering valuable insights into the country's efforts to balance international conventions and national legislation in the pursuit of environmental protection.

Chapter 8, "Forest Policy and Management in Colonial and Post-Colonial India: A Historical Study," authored by Prashant Tripathi, provides a historical perspective on forest policy and management in India, from the colonial era to post-independence. The author highlights that India, with its geographical diversity and various agro-climatic and vegetative zones, has a variety of cultural approaches to resource management. Forests have traditionally been man's habitat in India, with ancient Indians regarding forests as the mother and the queen, "Aranyani." The study of British colonial rule has brought attention to India's forest history.

In light of the ongoing discussion about the proposed Forest Act in India, an analysis of Indian forest policy has become necessary. The chapter focuses on the historical process of forest policy and management in India, discussing the colonization of Indian forests, colonial forest policy, the establishment of the forest department and management system in India, and the social, economic, and environmental impact of forest policy and management. The chapter also examines the relationship between the state and local communities affected by the forest policy system. Furthermore, it re-examines the legacy of forest policy and management in post-colonial India, the implementation of new ideas, and the challenges and outcomes. This chapter offers valuable insights into the evolution of forest policy and management in India and its impact on the environment and local communities.

Chapter 9, "Visual Language and Environmental Representation in Moroccan Cinema 'Sotto Voce' as a Case Study," co-authored by Mohamed Mliless, Latifa Belfakir, and Fouad Souiba, delves into the representation of the environment in Moroccan cinema, specifically focusing on the film *Sotto Voce* by Kamal (2014). The authors aim to identify how the environment is represented in the film and analyze whether it is portrayed as part of the location or as a means of discussing the main theme of the film. A close analysis of the visuals and narratives sheds light on how modern Moroccan cinema represents the environment and its role in shaping public understanding of the relationship between humans and the natural setting. The chapter attempts to deepen our understanding of cinema's role in shaping our perception of the environment by examining how the producer has used visuals to develop the film's theme. The study analyzes a war film that uses the environment as a setting for its theme, examining several visual elements as a language to portray nature. The authors' approach, based on visual analysis, allows for a close examination of thc visual elements to explain how the film represents the environment and conveys its themes and messages. Various forms of shots and scenes were identified that underline the beauty of the natural environment in relation to humans, creating a sense of intimacy or distance between characters and the natural context. The study reveals that the environment is a prevalent theme in the film, serving as a catalyst event to trigger human feelings and emotions. Interestingly, praise for the environment is linked to Deaf and Hard of Hearing (DHH) characters in the film. The results also show that characters, mainly the DHH, are often shown in nature, highlighting its beauty and significance. Additionally, dying people are often depicted in nature, suggesting a connection between the environment and the cycle of life. Despite the prominence of this imagery in the film, it does not provide any explicit environmental awareness tips or messages. One notable finding is that nature is portrayed as a source of reconciliation in the film. Characters are often shown finding peace and solace in nature, suggesting its restorative properties. One final finding from the film is that nature's importance is represented as a silenced voice, in contrast to the sounds of detonations that convey human emotions. In conclusion, cinema can play a vital role in exposing environmental issues and inspiring audiences to take action. While it is often associated with fun and entertainment, it can also serve as a powerful tool for social and political commentary.

Chapter 10, “The Representation of Hunted Animals on Social Networks: A Visual and Critical Discourse Analysis,” co-authored by Fouad Akki, Mohammed Larouz, Hicham Laabidi, and El Mehdi El Madani, discusses the portrayal of hunted animals on social networks, providing a visual and critical discourse analysis. The authors note that visual communication research has not sufficiently addressed how ecosystems are depicted in social media. Therefore, this study analyzes public discourse on images of hunted animals displayed on Facebook sites in different settings worldwide. Situated within visual environmental communication, this qualitative research uses frame analysis and a multimodal approach that combines semiotics and critical discourse analysis to analyze how audiences perceive the images of hunted animals. The comments of Facebook users on these visuals were analyzed ecolinguistically to determine their role in constructing messages on environmental issues.

The findings reveal that hunted animals are perceived as commodities, confrontations, and achievements. While most comments, primarily from hunters or people interested in hunting, support and encourage trophy hunting, a few comments from people outside this domain discourage and disagree with these killings and unethical acts. The study suggests showing enough public concern about environmental issues, exemplified by the problem of killing and displaying non-human creatures in visuals. It also recommends regulating human behavior to ensure ethical hunting and reduce the negative impact of poaching. This chapter provides a unique perspective on the portrayal of hunted animals on social networks and the role of visual communication in shaping public discourse on environmental issues.

Chapter 11, “The Rain Sound as an Affective Force that Unites Humans and Nature,” authored by Jaouad Achtitah, delves into the profound connection between humans and nature, embodied in the universal experience of listening to the rain. The author adopts the “affect theory,” which links our mind, body, and the world in different ways, to examine how the sound of rain impacts us. The “affect theory” provides a basis for examining the connection between the sound of rain as a natural phenomenon and human cognition and emotion. The author posits that since there is sufficient evidence that the human psyche is a natural phenomenon, the sound of rain could have an affective influence on us. This suggests that the sound of rain could affect our body and mind, thereby establishing a firm human-nature connection. The chapter argues that the affectiveness of the rain sound provides us with another dimension of connecting to the web of life. Accordingly, the author raises two fundamental, debatable questions: How does the rain sound affect the human psyche? How does the rain sound affect the human-nature relationship? This chapter offers a unique perspective on the sensory experiences and their impact on our relationship with the environment, particularly focusing on the sound of rain as an affective force that unites humans and nature.

The latter part of the book delves into the psychological and societal aspects of environmental interaction, discussing the role of the environment in shaping identity and Chapter 12, "The Environment between Psychology, Identity, and Society in Morocco as a Case Study," authored by Douae Serraf, is a transdisciplinary work grounded in a socio-environmental study. It explores the intersection of environmental studies with various disciplines such as education, law, politics, media, psychology, and translation. The author posits that the environment, initially associated mainly with nature, has been extended to various areas of study, revealing a more intersectional perspective. When environmental issues are viewed as human issues as well, it becomes clear how individuals' identities and behaviors could be significantly affected by ecological complexities. This chapter addresses the environment with three main areas of study: environmental psychology and identity. The author acknowledges that the intersection of these disciplines would explain and elaborate on nature's close connection with the universe and humanity. The study of human-environment interactions is a constantly evolving field of research with various attempts at change and resistance. The chapter studies the overlapping variables that the environment has with individuals, their identities, mental health, and the impacts on society. Douae Serraf suggests empirical research and theory on how identity relates human responses to the natural environment in Moroccan society. The chapter attempts to answer the following questions: (1) To what extent is the Moroccan identity reflected in the natural environment? (2) Can human psychology improve or deteriorate because of the environment in Morocco? And (3) to what extent does nature influence human behavior in Morocco? This chapter provides a unique perspective on the interplay between psychology, identity, and environmental issues in the context of Moroccan society.

Chapter 13, "Ecofeminism in Morocco between Theory and Reality," authored by Hafsa El Bastami, works towards finding solutions to eliminate the marginalization of women, an issue that has long been at the center of attention of international organizations, world leaders, and activists. The author asserts that for a society to progress and develop, gender inequality and all forms of discrimination against women must be eradicated. She highlights the importance of programs and projects that have been developed specifically for this purpose. Morocco is among the countries that have been working tirelessly to ensure gender equality, women's emancipation, and economic, social, and political development, addressing these issues enthusiastically at national, regional, and international levels. The chapter notes that a number of leading Moroccan organizations, created in the 1980s and 1990s, have been driven by feminist thinking and have set their missions and goals towards ensuring gender equality in all its aspects. However, the concept of ecofeminism, coined in 1974 by the French feminist Françoise d'Eaubonne, predates these organizations. This line of thought sought to focus on the intersection and interconnectivity of women and nature, highlighting their similarities, specifically when it comes to being discriminated against due to the dominance of patriarchy.

Through a selection of scholarly articles, this chapter introduces the theory of ecofeminism and how it is depicted in Morocco as a case study. The results show that even though there are numerous organizations in Morocco with a focus on protecting the environment, Morocco still has a long way to go to fully embed ecofeminism nuances in the work carried out by those organizations. This chapter provides a unique perspective on ecofeminism in Morocco, highlighting the ongoing efforts and challenges in achieving gender equality and environmental protection.

Chapter 14, "The Environmental Crisis, Postcolonialism, and Gender Inequality in the Global South: A Postcolonial Ecofeminism Approach," authored by Hajar Taha, explores the intersection of the environmental crisis, postcolonialism, and gender inequality in the Global South through a postcolonial ecofeminist approach. The author argues that the political, economic, and social dimensions, specifically regarding agricultural practices, are interconnected and rooted in power inequality and patriarchy. She hypothesizes that postcolonial ecofeminism is more adequate to translate the experiences of female farmers in the Global South. The chapter assesses how gender bias impacts female farmers' productivity in the Global South and the challenge that the male model of agriculture poses for development. It also examines the linkages between climate bias and nature management to argue for the improvement of women's project management, decision-making, and leadership capacities. As a methodology, the author uses postcolonial ecofeminism theory, focusing mainly on theoretical concepts such as the "male/female" dichotomies, "the feminization of agriculture," and "the invisibility of women's labour." She draws conclusions from well-documented studies to support her argument that women's farming approaches can be more sustainable. The chapter is conducted at three levels: global, regional, and local. At the global level, the focus is on the Global South; at the regional level, the focus narrows down to Africa; and at the local level, the focus is on Morocco. The author uses a level-based analytic structure to balance on-the-ground women's concrete experience drawn from field research with postcolonial ecofeminist concepts and broader studies of global and regional trends. This chapter provides a unique perspective on the environmental crisis, postcolonialism, and gender inequality in the Global South, offering valuable insights into the role of postcolonial ecofeminism in addressing these issues.

Chapter 15, "Women and Land Property within an Ecofeminist System," authored by Redouane Bougtib, explores the role of women in environmental conservation, providing a feminist perspective on environmental issues and highlighting the importance of gender equality in environmental conservation. The author examines the strategies that patriarchy and capitalism use to oppress women and the environment. The chapter argues that ecofeminism, as an intellectual and social movement, aims to create change in the lives of oppressed women through awareness-raising and education about the importance of justice between men and women and between humans and nature. This chapter addresses how ecofeminism strives to shed light on patriarchal practices against women and the environment at large through academia. It aims to examine the importance of doing justice and providing protection for both women and nature as two interdependent elements. The author conducted fieldwork in two separate locations: the region of the Oulmes Mountains and the Sidi Kacem (Bir Taleb) location. The findings reveal that natives in both territories have different perspectives on granting women access to collective lands and women's knowledge of their right to collective lands as right holders (sullaliyates). This chapter underscores the importance of ecofeminism in advocating for gender equality and environmental conservation, offering valuable insights into the intersection of women's rights, land property, and environmental conservation.

In Chapter 16 titled "The Two Cultures, the Third Culture, and the Environment in Ian McEwan's *Solar*," Nasreddine EL Guezar provides an insightful critique of the representation of environmental issues in literature. The chapter offers a comprehensive analysis of Ian McEwan's novel *Solar*, focusing on the intersection of science, literature, and environmental concerns. Amidst the prevalent schisms within academia, the author posits that literary fiction has extensively addressed this topic. McEwan's 2008 novel, *Solar*, serves as a relevant text to examine this perspective. The novel not only delineates profound academic divisions but also underscores potential avenues that bridge these divides and mitigate academic discord. These divisions and opportunities for connection warrant attention, as they play a pivotal role in the pursuit of holistic solutions to the environmental issue of climate change, as depicted in the novel. Through an ecocritical lens, and more specifically through the framework of C. P. Snow's prominent concepts of "the two cultures" and "the third culture," this chapter endeavors to elucidate these ideas. Solar by Ian McEwan portrays the inadequacy of scientists in addressing environmental issues. This is vividly dramatized through the protagonist, Michael Beard, whose resistance to collaboration with non-scientific domains, primarily the humanities, impedes his ability to adequately address climate change.

Chapter 17 of the book centers on "Ecotherapy and Communion with Nature in Farid Alansari's 'The Return of the Cavaliers': An Ecotheological Perspective." Actually, Chapter 17, authored by Abdelaziz Dqaqi, delves into the therapeutic benefits of nature, elucidating how communion with nature can enhance mental health and well-being. Farid Alansari, predominantly recognized as a religious scholar, has left an extensive cultural legacy, with his works spanning religious, political, and literary domains. However, he is more renowned as a religious scholar and political activist than a literary figure. The author contends that Al-Ansari's novels and poems have not been sufficiently scrutinized, with scant attention paid to the ecological discourse in his works. The chapter raises the question of whether literature with a religious backdrop can serve as a suitable medium for ecological discourse, given the principles proposed by ecotheologists to absolve religions of potential environmental transgressions. The chapter aims to explore this question by examining Farid al-Ansari's novel *The Return of the Cavaliers*, emphasizing two primary themes: ecotherapy and communion with nature. The novel, originally written in Arabic, was subsequently translated into English by Amany Shalaby.

1

Nature Deterioration

From Anthropocene to Capitalocene

Mohammed Adli and El Houcine Chougrani

Certainly, nature plays an important role in people's everyday life. No one can deny that nature always surrounds them; they interact with it now and then. The biosphere is the most crucial part of nature regarding the man-nature relationship. The reason behind this is that nature not only surrounds human beings, but all the other creatures on the earth as well. This fact pushes us to claim that human beings cannot live normal lives without nature. It can be even claimed that man is in need of nature more than the latter needs him since nature is able to exist without man, whereas humans can never exist without nature. As a matter of fact, man has always been depending on nature. Nature has represented and still represents, for him, the various sources of physical and psychological well-being. This leads us to confirm that as soon as nature continues to be man's source of well-being, a deep and strong relationship between him and nature will always remain intact and can never be separated. Generally, man's relationship with nature can be divided into three main stages. The first stage is when individuals used to rely entirely on nature because they consider it the source of happiness and well-being. The second stage is a bit more different, because man's selfishness and eagerness for profit led him to treat nature differently and forget his responsibility towards it. As for the last stage, it will be concerned with nature's indifference towards man's actions and attitudes. It should be said that the actual evolution of deterioration emerged from the huge constraint of the accumulation of the capitalist system together with the infinite economic growth determined by the greed of humans.

We use the analytic methodology to evaluate the process of deterioration and the level of responsibility of human beings, especially from the Industrial

Revolution era and the infinite accumulation of capitalism. In this context, it is imperative to demonstrate the contending process of deterioration.

Today, there is a fundamental question that seems to impose itself heavily about the survival of humanity against the greed of Capitalocene: Should we continue to enhance infinite economic growth? Or should we make a pause (la *décroissance*) to rethink our behavior? Sure, that linear economic process is not a solution, but rather a serious problem. What is worse is that even poor countries cannot enhance their environmental progress, because they have many priorities against climate vulnerability, which means that they need tremendous efforts to finance climate resilience.

Man and Nature: From Harmony to Human Containment

Man's Connection to Nature

Undoubtedly, man's relation with nature has always been very special. In the first stage, individuals used to rely entirely on nature. At that time, they seemed to be very happy to live in harmony with nature; they had a great deal and endeavor to protect animals and birds and to clean the rivers and lakes. In short, humans used to consider nature as a part of their lives; they felt glad to look at a tree and see its beauty. The reason behind this fact is that man was very conscious that establishing a close relationship with nature simply means that he will be responsible for it and for all the living creatures on the earth. In line with this idea, various researchers such as Nisbet et al. (2009) and Mayer and Frantz (2004), for instance, confirm that individuals, who are strongly connected to nature, tend to be more characterized with many positive characteristics such as being more conscientious, extraverted, agreeable, and open on life. This belief is based on the argument that the connection to nature can, definitely, be associated with more pro-environmental attitudes and a greater willingness to engage in sustainable actions. This is why we find them more likely to spend their time outdoors in nature and be engaged in a variety of pro-environmental behaviors.

In fact, man's close relationship with nature provided him with a broad sense of well-being. Muirhead (2011) and Dinnie et al. (2013), for instance, believe that well-being can help all human beings to feel good in their health and function very well and happily in their society. Of course, interacting with nature boosts people's self-esteem and their mood (Kuo & Sullivan, 2001). It also reduces feelings of anger (Moore et al., 2013), and even improves the general psychological well-being with positive impacts on emotions and behavior (Kaplan, 2001). One of the best examples in this context are Barton and Pretty (2010) who believe that taking exercise in nature can be more beneficial and better to relieve anxiety and depression than an exercise taken in urban settings. Even in terms of cognitive benefits, some researchers such as Tyrväinen et al. (2014) have also found that green spaces can be very restorative as they contribute to intentional recovery and reduce mental fatigue. This is why they insist that the exposure to both urban nature and wilderness settings will definitely yield to positive effects on academic performance and the ability to concentrate and perform mental challenges.

More significantly, several studies of adults' and children's physiological well-being have proved that time spent in natural settings can really relieve the negative physiological effects of various stressors, especially in urban areas. In an attempt to confirm this idea, Aerts et al. (2018) believe that indirect physiological health benefits are derived when nature promotes physical activity. Therefore, they see that the exposure to greenness can be associated with lower probabilities of many health problems. More importantly, it is believed that natural environments and shared nature experiences can have a very beneficial effect on health by providing opportunities for social interaction and strengthening bonds within families and friends (Greenwood & Gatersleben, 2016), because nature also fosters both personal and social formation.

Since man's relationship with nature has always been an important topic for discussion, it is of paramount importance to associate this relation with happiness. Various researchers such as Peterson et al. (2005) list three important elements to reach happiness, which are: pleasure, meaning and engagement. Undoubtedly, theses three principles control the relationship between man and nature and allow human beings to experience happiness provided that there is a sense of commitment on the part of man. The latter, indeed, will be very pleasant and satisfied in nature, because it provides him with all his necessities and desires: a clean air, water, food, etc. In this context, Capaldi et al. (2014) concluded that happiness appears to be associated with a strong personal connection with nature. This connection becomes stronger when individuals see nature as integral to them, rather than separate from them (Pritchard et al., 2020). Man is also pleasant because he understands that nature is very important for his existence and for his well-being on condition that he will sign a contract of being responsible for its protection from any threat or danger. However, it must be declared that in case man does not respect his commitment towards nature, the consequences may be worse, as we are going to see in the coming chapters.

Man Against Nature

As stated earlier, the second stage is a bit more different because man did not respect his commitment towards nature. Man started to ignore his responsibility towards the protection of nature. The technological advancements and man's eagerness to modernity and civilization pushed him to destroy nature willingly or unwillingly. Nature, undoubtedly, has its laws designed to maintain balance. For instance, if the number of any living species tends to increase out of proportion, some force will arise to control it. There is equilibrium in undisturbed nature between food and feeder, hunter and prey, so that the resources of the earth are always maintained. Some animals or plains may seem to dominate the rest, but this can happen only as long as the general balance exists. It should be noted that these laws can never be disregarded without various negative consequences.

Certainly, nature, which surrounds human beings and seems to influence their existence on earth, cannot easily accept the total ignorance of its laws as an excuse to legitimize other purposes. Nature imposes its laws and warns human beings that if they do not carry out its orders, the results will be very terrible. This means that if humans want to guarantee their existence and continuity, they have to take its warnings in a serious and responsible way. Today, there is no wonder that man is constantly aware of nature's influence around him be it in the air he breathes, the water he drinks, the food he eats or in the flow of energy and information they uses. More importantly, individuals are aware that many of his troubles are a response to the natural processes and changes in the weather, intensified irradiation of cosmic energy and the magnetic storms that rage around the earth. In short, man is conscious that he is strongly connected to nature and that he can never live outside it.

At the present moment, it seems that man and nature have started to lose the sense of connection and commitment. As society develops, man tends to become less dependent on nature directly, while, in fact, indirectly his dependence is growing more and more and his power over it is progressively increasing. Under the pretext of modernity and civilization, man started to ignore the importance of nature and he no longer assumes his responsibility over its protection. One of the best examples that can be provided in this context are forests, which were destroyed. Also, several places where man had made his home and where the earth was cultivated have been badly and terribly exploited. Another example is related to eliminating industrial waste, which is also becoming increasingly complex. In addition, the threat of a global ecological crisis hangs together with the polluted air have all led to the crisis of climate change that the whole world is suffering from nowadays.

Today, one can safely claim that the previous dynamic balance, which used to exist between man and nature, is no longer running well on the right path. This is due to various annoying problems, among which we can speak about human's strong connection to technology. Surely, technology is playing a vital role in man's life and it can be safely mentioned that individuals are ready to do anything for the sake of preserving and maintaining their attachment to it, simply because they cannot imagine how life would be without technology. The capitalist system shows that it very difficult today to satisfy the needs of human beings and society in general. A human action against the laws of nature and against the harmony of the biosphere seems to threaten all humanity. Bearing all these in mind, one would question the fact whether or not there could be, in the near future, any fear of a universal disaster that may soon arise and threaten the whole world. In an attempt to clarify this fact, it would be of great significance to make reference to what Malthus (1798), who focused in his Malthusian theory, on the strong relation between population growth and food supply. Malthus proposed a theory of population, where he explained that the population is growing exponentially while the food supply is growing arithmetically. This is why a balance between the two should be established through positive and preventive checks. The Malthusian theory explained that the human population is growing more rapidly than the food supply and, unfortunately, this growth is not going hand in hand with the earth resources. For him, the human population has risen over the past three centuries.

Nature Indifference

Undoubtedly, nature is declining at an unprecedented speed. A huge number of species all over the globe are facing extinction. More importantly, the excessive use of technology tools, temperature, pollution, waste, population overgrowth, and nuclear experiments, for instance, are all cases that tend to destroy nature at a very high speed. Every year, these problems cause millions of premature deaths. This implies that almost one third of all rivers in Latin America, Asia and Africa suffer from severe pollution, which explains that nature's capacity to support humans started to be plummeted. Air and water quality are reducing, soils are depleting, crops are short of pollinators, and coasts seem to be less protected from storms.

The industrialization process seems to have alienated man from nature. Generally, as the human population continued to soar, coupled with technological advancements, man started using more efficient methods of sustaining himself. All societies have evolved and the demand for resources shot drastically. This fact has pushed humans to distance themselves from nature. In other words, man is no longer in harmonious relation with nature and, consequently, an imbalance status between him and nature has arisen because of the excessive treatment and exploitation. Of course, no one can deny that the rise of industrialization has also contributed to man's gradual alienation from nature.

More importantly, the global world today is witnessing serious environmental problems related to climate change. Many countries, especially very developed ones like The United States and China, for instance, are considered responsible for this unprecedented crisis. Today, it is estimated that Greenhouse-gas emissions (GHG), have reached their highest-ever level in 2021 with global carbon dioxide emissions from fossil fuels. China is, currently, the highest emitter followed by the Unites States and Russia. This climate change is, no wonder, having devastating impacts on the lives of human beings. It should be stated that the effects of climate change may range not only from sea level rise or adverse weather conditions, but also to disease outbreaks such as the corona virus.

Generally, it has become very evident that no country will be safe from the consequences of climate change. Fossil fuel firms directed by developed countries, which do not wish to reduce or stop their air pollution, will also be victimized. Of course, either the US or China together with other countries are competing each other; each one of them wants to tell the world that it is the most powerful and developed one even at the expense of climate change and nature. In other words, the capitalist world does not acknowledge emotions or feelings of mercy and compassion towards the living creatures or nature. For them, stopping or even making a pause means that they will no longer be powerful and that other countries will take their position. Of course, this fact can never be appropriate or accepted by the capitalist competitive countries, whose mail goal is to achieve profits and preserve their interests regardless of the means or damages.

However, there is an important question that seems to impose itself heavily in our everyday context. What is the reaction of developing or even poor countries towards climate change and the terrible actions of developed countries towards nature? Do these people have the right to live in peace and harmony with nature? Or should they remain as followers who need only to carry out the orders of developed countries?

In fact, life is a divine right that is given by God to all creatures, including human beings; all of them have the right to enjoy life and live in peace under the umbrella of nature. However, due to the selfishness of a category of humans, whose main goal is to achieve their profits and preserve their interests, a large number of creatures, including human beings, will pay a heavy bill. Uncertain future is looking due to mysterious processes such as water vapor and ice feedbacks, ocean circulation change, and natural cycles of greenhouse gases. These processes represent the source of human destruction. Indeed, the extent of future climate change depends on what humans do now to reduce greenhouse gas emissions. The more developed countries emit the larger future changes will be.

European and American Environmental Containment Policy

European and American Environmental History: Industrial Era and Anthropocene Epoch

The Anthropocene is a contested term, coined by Crutzen (2002) to describe the contemporary geological era, which begins once humans became a planet-changing (Lorimer & Driessen, 2014). This change contributed to new relations with nature. The term Anthropocene, introduced about a decade ago, has recently gained momentum, but informally, used in the global change research community. However, the term has yet to be accepted formally as a new geological epoch or era in Earth history (Will et al., 2011). This new period of nature domination became a fact to destroy the natural resources. Many geologists are, now, tracing the first stage of the Anthropocene back to the industrial revolution in the nineteenth century, and delineating a second stage with the first atomic explosions in the 1940s (Rousell, 2016). The industrial revolution is a turning point to change the paradigm to create a need and to consume more and more.

So, when did the Anthropocene actually start? In fact, it seems to be difficult to put an exact date on the transition that occurred at different times and rates in different places, but it is clear that in 1750, the Industrial Revolution had barely begun; it had completely transformed England and spread to many other countries in Europe and across the Atlantic to North (Steffen et al., 2011). The evolution of man vision to nature considered a critical point to destroy the nature. The term Anthropocene suggests two aspects: the first one is that the Earth is, now, moving out of its current geological epoch, called Holocene and as for the second aspect, it means that human activity is largely responsible for this exit from Holocene, that is to say, that humankind has become a global geological force in right. Since its introduction, the term Anthropocene has become widely accepted in the global change research community (Steffen et al., 2011). The Anthropocene is the key word to determine the man responsibility.

However, in the beginning of the twentieth century, nobody, except perhaps Vernadsky in the USSR and Henry Adams in the United States, imagined the Great Acceleration of the second phase of the Anthropocene: the post-Second World War worldwide industrialization, techno-scientific development, nuclear arms race, population explosion and rapid economic growth (Steffen et al., 2011). These huge transformations of the occidental society have become a decisive element to kill the nature resources. Crutzen and Stormer (2000) date the Anthropocene to the industrial era: they propose the latter part of the eighteenth century, although some may even want to include the entire Holocene effects of human activities, which have become clearly noticeable. This is the period when data retrieved from glacial ice cores, showing the beginning of a growth in the atmospheric concentrations of several "greenhouse gases," in particular C02 and CH4. Such coincides with James Watts's invention of the steam engine in 1784 (Braje, 2015). Human history mentions the huge transformation of nature to industrial benefit.

The naming of the industrial era as the beginning of the Anthropocene Age is a scientific periodization that brings together the analytical and experimental narratives of planetary sustainability and historical narratives of industrialization and expansion (Kavita, 2014).

The Third World Nations: Why not responsible? As they became known as "developing nations," and later as "emerging economies," the industrial practices and waste products that went along with development were to become an issue in climate talks. While there was a range of "Third World" governmental responses to climate change, there was a common thread in demands for the industrialized North to take the lead in addressing the pollution effects and resource depletion that the West's Industrial Revolution and colonial expansion had set in motion (Kavita, 2014). The common, but differentiated responsibility is not a principle to protect nature. It is not the third world's responsibility.

A variety of scientific methods have been employed to measure the magnitude of planetary changes associated with the Anthropocene, including sea level rises, global temperature increases, levels of atmospheric carbon dioxide, rates of anthropogenic denudation, and human population growth (Rousell, 2016).

Why the Tragedy of the commons? The Anthropocene, as Crutzen and Stoermer argue, is the era when human impacts have shaped geologic time. Although not accepted universally as a scientific term, the Anthropocene has gained much cultural and political resonance in the last five years. In his piece, "The Great Acceleration and the Great Divergence: Vulnerability in the Anthropocene," literary scholar Rob Nixon suggests that although environmental crises and human inequality are deeply intertwined, narratives about the Anthropocene have generally "sidestepped the question of unequal human agency, unequal human impacts, and unequal human vulnerabilities" (Sze, 2015).

In fact, there is no question about humanity's contribution to global warming, species loss, ocean acidification, extreme weather events, rising sea levels, decreasing ice, and retreating glaciers. However, the scale of possible actions that human beings can take to slow, stop, or reverse other effects of climate change (or, indeed, to stop climate change itself) is dubious at best. The simple reality is that we seem to be on an exponential trajectory, and every year things get much, much worse, despite the increasingly frenzied rhetoric. There are many reasons for this: to talk about climate change without talking about the meat industry, for instance, seems hollow, and green activism must involve detailed and consistent references to critical animal studies (Simon, 2018). We cannot ignore the interdisciplinary studies to determine human responsibility.

Many scientists suggest that today, we are living in a new era of natural history—the Anthropocene—where human activities have introduced new biophysical factors into the biosphere. More importantly, these activities have begun to change the physical parameters that determine the functioning of the major earth system processes, most noticeably, the composition of the atmosphere (Simon, 2007). It should be mentioned that this transformation is enhancing a significant negative transformation.

Interdisciplinary Science Versus the Anthropocene

The Anthropocene incorporates interdisciplinary scientific analysis and evaluation of the human impact on global biology, geology, and ecology, which are, thus, inseparably connected to cultural developments (Kersten, 2013). The Anthropocene not only encapsulates various interconnected ecological issues and concerns both globally and locally "under a single heading," but also more radically, reinforces the observation that human society is symbiotically embedded within its ecological and technological situation (Machin, 2019). Of course, the connection between these entire scientific and social fields can help to resolve the problem.

How can one survive in the Anthropocene Age? First of all, the Anthropocene is the earth time of humans. As for the second aspect, the Anthropocene describes a space dominated by humans: Humans see and shape the Earth as "their" world (. . .). Concerning the third aspect, the central actor in the Anthropocene is humanity, which is altering the Earth's ecosystems with dramatic consequences for the environment that humans need for their survival. The fourth aspect is related to the concept itself, which is a product of human thought and reflection based on: Who else can name, describe, and manage this new Earth period, if not human science? (Kersten, 2013).

The Anthropocene concept is not only about facts but also automatically about ethics and laws as well; not only about science but also about politics and governance (Kersten, 2013). So, ethics studies should be integrated to contribute to the solution of the effect of the Anthropocene. Perhaps, the decolonial Anthropocene Visuality can be seen as a clear solution, the Industrial Revolution represented not so much a shift in energy use as the emergence of a new knowledge economy. In this model, technological innovation was the fruit of intellectual exchange and the Baconian ambition to conquer nature rather than a direct response to an ecological crisis or pressing scarcity (Albritton, 2012).

In this context, Srinivas advocates for scholars and intellectuals to assume responsibility for reconfiguring contemporary knowledge production processes, deploying their specialized competencies within the broader framework of Enlightenment epistemology to confront the most pressing and multifaceted intellectual, social, political, ethical, and scientific challenges of our era, particularly those pertaining to geological and climatic transformations, as well as the emergent conceptualizations of temporality and historical understanding that such awareness has engendered (Brown, 2016).

We know that life is characterized by "capriciousness, temerity, opportunity, and chance" (Vico, De anti quíssimaital forums sapientia). This is the problematic concept of the Anthropocene (Brown, 2016). We can overcome this complex problem. The Anthropocene was in the past, is at present, and will be in the future a period not of harmony created by institutions, but of severe global conflicts of national and economic interests for resources, driven by social welfare, market competition, and—of course—by political sovereignty (Kersten, 2013).

'The global environmental problems that the Anthropocene encapsulates, highlights, and connects (ozone depletion, acid rain, biodiversity depletion, ocean acidification, climate change, etc.) do not conform to conventional political boundaries. Just as the causes of these various problems cannot be located within any single territorial state, the solutions cannot be generated by them acting alone without coordination. The Anthropocene renders state boundaries as simultaneously anachronistic and obstructive (Machin, 2019).

The collapse of any credible distinction between humans and nature forces humanity to modify ethical codes or political aspirations to no longer rely on simple models of autonomy and separation. The historically acceptable assumptions about the earth as the given home for humanity no longer hold. Many of humanity's changes to the biosphere have yet to show dramatic manifestations, but they are coming as a consequence of past human activities. They will not necessarily be malign or benign, but they will likely be unpredictable, and their consequences will be unevenly felt by people depending on the particular context in which they live (Simon, 2007).

For global security, some measures should be taken among which we can speak about: first of all, reducing the total ecological disruption of the global system by making energy use and material consumption in the global economy more efficient. The second measure is a security policy that must enhance the abilities of states and military forces to provide aid and assistance at short notice so that they can protect people and livelihoods when disasters happen. As for the third measure, it has to do with a logical extension from the second point; this policy must extend the habits and institutions of international cooperation so that aid and trade—rather than confrontation and conflict—would be the responses to ecological disruptions (Simon, 2007).

Growing concern over human vulnerability corresponds to the shift in the earth sciences over the last decade. Science has enabled us to recognize that we are living in a single and interconnected biosphere; one that humanity is changing as it becomes an urban species increasingly powered by fossil fuels.

Earlier, when environmental concerns regarding the dangers of industrial pollution existed, the need for resource and wildlife conservation have become a necessity. Even Malthusian concerns about population are being overtaken by a global perspective that emphasizes the ecological connections among phenomena and between humanity and its environment (Simon, 2007).

The story's narrator, a twenty-fourth-century Chinese historian, recounts that democratic governments were "at first unwilling and then unable to deal" with the "Great Collapse" caused by climate change, leaving only authoritarian states as survivors (Schneider & Mayerson, 2017).

Anthropocene/Capitalocene Are "Ecocide"

It is clear that if the Anthropocene implies ecocide and the current global regimes accelerate the process through tantalization trances, then there would be, even for the most peace-seeking people, an implicit war within the global disposition (Cohen, 2012).

The geologists who advocate the concept of the Anthropocene generally agree that if the effects of human activity, especially in the last two hundred years, are evident in a layer of earth's sediment, a new geological era should be designated to reflect humanity's footprint (Kennedy, 2012). A close study of the impact of humans on the planet and its inhabitants should be conducted. What is the connection between Neoliberalism and the Anthropocene? Nation states have given way to global corporate rule; especially biotechnological progress and capitalist consumption that have had a major influence on society and the environment (Schmeink, 2016). Also, questions of self-regulation tend to dominate accounts of the relationships between neoliberalism and the Anthropocene, whether in the form of the critique of capitalism as committed to unregulatable growth, or in the hope that capitalism can be transformed into a harmonious, self-regulating and ecologically neutral system through carbon-offsetting mechanisms, or through new forms of artificial intelligence, which seek to channel population-level differences algorithmically in "smart electrical grids" or "smart cities" (Mitchelln, 2021). It should be stated that both Margaret Atwood and Paolo Bacigalupi imagine a world of ecological disaster that would be caused by the human inability to see the consequences of its hypercapitalist endeavors together with the environmental changes enhanced by the Anthropocene (Schmeink, 2016).

The need to achieve effective planetary stewardship is urgent. As we go further into the Anthropocene, there is a risk of driving the Earth System onto a trajectory toward more hostile states from which we cannot easily return (Will et al., 2011). For some, the Anthropocene signals a final enclosure of politics and culture within ecology; new geopolitics in which Earth is the sovereign authority and humans are inmates of a planet-sized camp in a permanent state of emergency. Whereas for others, it is an occasion to double down on techno-hubris and call forth more fevered bouts of rationality and management. Optimistic commentators hope that naming this new epoch might accelerate action on the pressing challenges of our time—that the Anthropocene is an "unprecedented opportunity" a wake-up call for "planetary stewardship," or just good to think with (Franklin, 2015). Thinking about our survival, in this age, is a real critical question for humanity.

Contrary to the Anthropocene theory, which suggests that we are all responsible for the climate crisis, it is the capitalist system and its class champions that are responsible for the climate crisis. A system that has produced a systemic problem cannot, in fact, solve the problem, given that this is a carbon-based capitalist civilization. It seems that nothing short of the fundamental decarburization of production, consumption, finance, and every life world on this planet will save human and non-human nature. The Paris Agreement falls short of this imperative (Vishwas, 2018). The sanction is a very important element in addition to the ethical element, which is a vital aspect to saving the planet. The Anthropocene discourse is a confirmation of the logic of imperial ecocide, having its genesis in the origins of capitalism and its making over 500 years. It is in this process that capitalism is also constituted as a geological force shaping and determining the conditions of life forms and worlds. Moreover, it is argued that the Anthropocene embodies power relations that affirm racialized dominance as part of the imperial ecocide (Vishwas, 2018). The latter can make it to improve human behavior against the capitalocene, and its negative influence on the processes of natural deterioration.

Conclusion

Generally, several concepts such as coexistence, cooperation, harmony, or conflict, for example, have had a lot of influences on the connection and conflict between humans and nature. It is true that there have been three essential phases for this connection or conflict; the first is related to man's connection to nature, the second is concerned with man being against nature, whereas the last phase has to do with nature's indifference. The question, today, is about what we can do in this tragic and unprecedented crisis evolution. The Anthropocene is a decisive epoch that means the domination of nature by the accumulation of capitalism. Humanity, indeed, has realized some economic progress, but without taking into account the need to save nature and humanity. Of course, nature seems to be fragile in this context.

However, saving nature is not the preoccupation of a weak category of individuals; we are living in the age of the Capitalocene, which means the deterioration of nature not by people of Africa, Asia, or Latin America, but by westerners who accumulate wealth and industrial progress. Deteriorating nature is equal to deteriorating the commons, which implies that it would be completely impossible to maintain human survival in this context. Perhaps, the acceleration of the Anthropocene is a fact, so *de facto*, showing that we will all suffer. The developing and developed countries will surely suffer, but the resilience to this huge transformation of climate will be decisive. Undoubtedly, Capitalocene is "'ecocide," but the problem is when we can use litigation to end the tragedy. Civil society is a core element to modify the equation, but in the context of economic crises sustain, we will not be able to make a significant change. Today, is seems that the capitalocene is accelerating a linear economic process, but without any clear social progress

References

Aerts, R., Honnay, O., & Van Nieuwenhuyse, A. (2018). Biodiversity and human health: Mechanisms and evidence of the positive health effects of diversity in nature and green spaces. *British Medical Bulletin*, 127(1), 5–22. https://doi.org/10.1093/bmb/ldy021 (pubmed.ncbi.nlm.nih.gov)

Albritton, F. J. (2012). The Industrial Revolution in the Anthropocene. *Journal of Modern History, 84*, 676–696.

Barton, J., & Pretty, J. (2010). What is the best dose of nature and green exercise for improving mental health? A multi-study analysis. *Environmental Science & Technology*, 44(10), 3947–3955. https://doi.org/10.1021/es903183r

Braje, T. (2015). Earth systems, human agency, and the Anthropocene: Planet earth in the human age. *Journal of Archaeological Research*, *23*(4), 1–31. https://www.jstor.org/stable/43956793

Brown, L. (2016). From Tropicopolitan to Anthropocene. *Eighteenth-Century Studies*, 50(1), 101–104. https://doi.org/10.1353/ecs.2016.0043
Capaldi, C. A., Dopko, R. L., & Zelenski, J. M. (2014). The relationship between nature connectedness and happiness: A meta-analysis. *Frontiers in Psychology*. Retrieved from https://doiorg.georgefox.idm.oclc.org/10.3389/fpsyg.2014.00976
Cohen, T. (2012). Polemos 'I am at war with myself' or, Deconstruction™ in the Anthropocene? *Oxford Literary Review*, *34*(2), 239–257.
Crutzen, P. J. (2002). Geology of mankind. *Nature*, 415(6867), 23. https://doi.org/10.1038/415023a
Crutzen, P. J., & Stoermer, E. F. (2000). *The "Anthropocene"*. IGBP Global Change Newsletter, 41, 17–18.
Dinnie, E., Katrina, M., & Sue, M. (2013). 'Reprint of Community, cooperation and conflict: Negotiating the social well-being benefits of urban green space experiences: Landscape and urban planning. *Journal of Land Scape and Urban Planning, 118*, 103–111. Retrieved from https://doi.org/10.1016/j.landurbplan.2012.12.012
Franklin, G. (2015). When horses won't eat: Apocalypse and the Anthropocene. *Annals of the Association of American Geographers*, *105*(2), 351–359. https://doi.org/10.1080/00045608.2014.988100
Greenwood, A., & Gatersleben, B. (2016). Let's go outside! Environmental restoration amongst adolescents and the impact of friends and phones. *Journal of Environmental Psychology*, *48*, 131–139. https://doi.org/10.1016/j.jenvp.2016.09.007
Kaplan, S. (2001). Meditation, restoration, and the management of mental fatigue. *Environment and Behavior*, 33(4), 480–506. https://doi.org/10.1177/00139160121973106
Kavita, P. (2014). Doing interdisciplinary Asian studies in the age of the Anthropocene. *Journal of Asian Studies*, *73*(4), 975–987. https://doi.org/10.1017/S0021911814001648
Kennedy, R. (2012). Humanity's footprint: Reading "rings of Saturn" and "Palestinian Walks" in an Anthropocene era. *'Biography*, *35*(1), 170–189.
Kersten, J. (2013). The enjoyment of complexity: A new political anthropology for the Anthropocene? *Perspective Issue*. https://doi.org/10.5282/rcc/6213
Kuo, F. E., & Sullivan. W, C. (2001). Environment and crime in the inner city: Does vegetation reduce crime? *Journal of Environment and Behavior, 33*(3), 343–367.
Lorimer, J., & Driessen, C. (2014). Wild experiments at the Oostvaarders plassen: Rethinking environmentalism in the Anthropocene. *Transactions of the Institute of British Geographers*, *39*(2), 169–181. https://doi.org/10.1111/tran.12030
Machin, A. (2019). Agony and the anthropos. *Journal of Nature and Culture*, *14*(1). https://doi.org/10.3167/nc.2019.140101
Malthus, T. R. (1798). *An essay on the principle of population, as it affects the future improvement of society, with remarks on the speculations of Mr. Godwin, M. Condorcet, and other writers. J. Johnson.*

Mayer, F., & Frantz, C. M. (2004). The connectedness to nature scale: A measure of individuals' feeling in community with nature. *Journal of EnvironmentalPsychology, 24*(4), 503–515. https://doi.org/10.1016/j.jenvp.2004.10.001

Mitchelln, R. (2021). Life, self-regulation, and the liberal imagination. In *Infectious Liberty: Biopolitics between Romanticism and Liberalism* (pp. 1st ed., 186–230). Fordham University Press. Retrieved from https://doi.org/10.1515/9780823294619-008

Moore, M., Shin. M., & Bruning, A. (2013). Cannabis use and risk of psychotic or affective mental health outcomes: A systematic review. *National Library of Medicine, 370*, 319–328. https://doi.org/10.1016/S0140-6736(07)61162-3

Muirhead, S. (2011). *Nature and well-being: Building social and emotional capital through environmental volunteering*. Scotland University of Dundee.

Nisbet, E. K., Zelenski, J. M., & Murphy, S. A. (2009). The nature relatedness scale: Linking individuals' connection with nature to environmental concern and behavior.' *Journal of Environment and Behavior, 41*(5). Retrieved from https://doi.org/10.1177/001391650831874

Peterson, C., Park, N., & Seligman, M. E. P. (2005). Orientations to happiness and life satisfaction: The full life versus the empty life. *Journal Happiness Studies, 6*, 25–41. https://doi.org/10.1007/s10902-004-1278-z

Pritchard, A., Richardson, M., Sheffield, D. et al. (2020) The relationship between nature connectedness and eudaimonic well-being: A meta-analysis. *Journal of Happiness Studies, 2*, 1145–1167. https://doi.org/10.1007/s10902-019-00118-6

Rousell, D. (2016). Dwelling in the Anthropocene. *Australian Journal of Environmental Education, 32*(2), 137–153.

Schmeink, L. (2016). *The Anthropocene, the posthuman, and the animal.* Liverpool University Press.

Schneider-M, M. (2017). Some Islands will rise: Singapore in the Anthropocene. *Resilience. A Journal of the Environmental Humanities, 4*(2). Retrieved from muse.jhu.edu/article/669724.10.1111/tran.12030

Schneider-Mayerson, M. (2017). Climate change fiction. In R. G. Smith (Ed.), *American Literature in Transition*, 2000–2010 (pp. 309–321). Cambridge University Press. https://doi.org/10.1017/9781316569290.021

Simon C. E. (2018). Hollow ecology and Anthropocene scales of measurement. *Mosaic: An Interdisciplinary Critical Journal, 51*(3), 37–52. https://doi.org/10027-1276-07/037016$02.00

Simon, D. (2007). Anthropocene geopolitics: Globalization, empire, environment and critique. *Geography Compass, 1*(1), 103–118. Retrieved from:https://doi.org/10.1111/j.1749-8198.2007.00007.x

Steffen, W., Grinevald, J., & Crutzen, P. (2011). The Anthropocene: Conceptual and historical perspectives. *Philosophical Transactions of the Royal Society A:*

Mathematical, Physical and Engineering Sciences, *369*(1938), 842–867. https://doi.org/10.1098/rsta.2010.0327

Sze, J. (2015). Environmental justice Anthropocene aarratives: Sweet art, recognition, and representation. *Resilience: Journal of the Environmental Humanities*, *2*(2), 103–118.

Tyrväinen, L., Ojala, A., Korpela, K., Lanki, T., Tsunetsugu, Y., & Kagawa, T. (2014). The influence of urban green environments on stress relief measures: A field experiment. *Journal of Environmental Psychology*, *38*, 1–9. https://doi.org/10.1016/j.jenvp.2013.12.005

Vishwas, S. (2018). *The climate crisis and systemic alternatives.* Wits University Press.

Will, S., Persson, A., Deutsch, L., Zalasiewicz, J., Williams, M., Richardson, K., Crumley, C., Crutzen, P., Folke, C., Gordon, L., Molina, M., Ramanathan, V., Rockström, J., Scheffer, M., Schellnhuber, H. J., & Svedin, U. (2011). The Anthropocene: From global change to Planetary Stewardship. *Journal of Ambio*, *40*(7), 739–761. https://doi.org/10.1007/s13280-011-0185-x

2

Food Production and Animal Welfare

The Stories We Are Enforced By

Diego L. Forte

Government and the Legal Forms

Considered as a discursive practice, laws, legislation, government directives, and texts produced in the different levels and stages of modern governments in general have an impact on our lives that is somehow different from the impact that other discursive practices have. They constitute an institutionally legitimized discourse intended to be "ideology free" in their regulation of behaviors (Vilajosana, 2010, p. 11). These discourses—legal forms in terms of Foucault—define subjectivities, knowledge, and power relations, and that is why they need to be analyzed (Foucault, 1978). From a systemic-functional linguistics perspective, legal forms are choices made by the users of the system, as any other discursive choices. But these choices not only constrain our social life by convention, they also enforce social resources that limit behavior in other ways; they are connected to the material power of society. In this sense, legal discourses are more than just part of the story; they may carry the same representations as other texts, but they also allow dominant discourses to enforce material measures, as well as impose certain ways of behavior and forbid others.

Legal forms possess vision over the practices and processes they regulate; they define subjectivities, types of knowledge, and power relations according to the particular—dominant—vision of the society (Foucault, 1978). In the case of animals, nature, and environment, they are subject to different kinds of regulations: environmental, commercial, alimentary, architectural, and wildlife regulations contain representations from dominant discourses in society and their legal status grants them the rank of "enforcement discourses" controlled by governmental, legislative, juridical representatives with the power to punish transgressions.

In Argentina, official controls over food production are carried out by the National Service for Food Health and Quality (Servicio Nacional de Sanidad y Calidad Agroalimentaria), known as "SENASA," its initials in Spanish. According to the Argentine government web page, the origins of the agency can be traced back to the nineteenth century, when the country needed to provide sanitary guarantees for its exports. Until the year 1996, SENASA stood for Servicio Nacional de Sanidad Animal (National Service for Animal Health). Back then, the agency devoted its duties to animal-origin food only. In that year, it was merged with the National Institute for Vegetal Sanity and Quality (Instituto Argentino de Sanidad y Calidad Vegetal, commonly known as IASCAV, for its initials in Spanish), conforming the current agency.

According to the web page of the Argentine government, the mission of the agency is to guarantee the safety of animal and vegetal origin foods through correct and adequate production processes. The agency plans, organizes, and executes specific programs to regulate production, directing it towards obtaining safe food for human and animal consumption (SENASA, 2020).

Although the change in the name of the agency evidence a neo-liberal imprint that imposes a commercial perspective over a sanitary one, that is, quality over sanity, its origins and mission were always related to animal and soil exploitation as the main industry of the country (Forte, 2019). As other commercial guaranties, Animal Welfare policies are a requirement for European customers, and they are therefore implemented by the agency to fulfill those requirements (European Commission, 2020). But as a technical agency, SENASA's legal discourses also involve elements of scientific and commercial discourses.

Objectives and Theoretical Framework

This chapter is part of a research that seeks to understand the way in which the discursive practices carried out by the agents of the Argentine Agency of Food Safety(SENASA) impact the social practices developed by the agency in the enforcement of legal discourses. Our aim is to analyze the relationship between discourses on animals and relevant professional practices—that involve material measures—of which they are the object. In this case, we focus on the Animal Welfare Manual, elaborated by the personnel of the agency—veterinarians—during the period 2015–16. This paper presents an analysis of the preface—where the organizational perspective is presented—and the images—that constitute a visual correlate of that organizational perspective. Our main theoretical framework is Multimodal Critical Discourse Analysis. We accordingly adopt the analytical tools proposed by Hodge and Kress (1993), Lakoff (2002), and Kress and van Leeuwen (2005). The preface presents the main framing of the information contained in the book and, as introductory texts, circumscribes the expectations of the readers to certain topics, approached in a specific manner. It particularly gives some clues about the way the manual is going to develop its contents. On the other hand, the images articulate the visual mode in accordance with the clues introduced in the preface.

The text is written in Spanish. Since the manual is prepared for the development of specific tasks, the interest that guides this research is to analyze how this technical information is framed and how it addresses its audience. According to Raiter (2003), dominant discourses act as an axis of references, structuring the most accepted version of social representations in a given society and putting away the less prototypical. In this sense, Flax (2020) points out that the dominant scientific discourse constitutes a certain kind of representation, keeping others outside, marginalizing them, and relegating them to the place of false or wrong thinking. Our working hypothesis is that the combination of scientific and neoclassical economic discourses structures the semiotic resources of the manual to avoid explicit mention of human agency in processes that negatively impact non-humans and the environment.

Science Economics and Animal Welfare

In the first place, it must be considered that the manual does not constitute a norm, law, or directive *per se*, but the instructions contained in it are mandatory for the official personnel of the agency. Therefore, it is a discourse that regulates a specific social practice and it is presented as a tool to apply scientific concepts to the management of farmed animals in the different stages of the food production. This locates dominant scientific discourse as one of the main sources of the text.

According to Verón (1987, p. 22), "scientificity" is an effect that certain texts produce in their meaning by problematizing their conditions of production.

Based on this definition, the manual does not qualify as scientific discourse; it describes how to apply concepts and techniques to a social practice; although it claims to be based on the latest scientific and technological investigations in the field (SENASA, 2015, p. XV). In addition, the categories that structure the narrative are provided by commercial-neoclassical-economics discourse: *nature is a resource, more is better than less, economic growth is good* (Stibbe, 2015, p 35). These representations orient the disposition of scientific discourse through the whole text.

Framing Animal Welfare

According to Coulthard et al. (2017), the legal style is determined by its history and function. What is being done has a direct impact on word choice (both lexical words and functional words), syntax and punctuation, and this, in turn, produces the distinctive register that can be measured and observed in frequency lists and legal text and talk. These choices are a direct consequence of the communicative purposes of legal language and the activities in which it is used. Therefore, we can say that lexical pieces found in official texts are meant to abide by certain practices. Holmes (2008, p. 9), describes the linguistic choices (who, where, what and why) made in any speech situation signaling the following categories:

- Participants: who is speaking and who are they speaking to? In our case, the manual is intended to SENASA's agents working in farming fields and slaughterhouses, but also to the general public, since official documents establish the legal standard for the activities they regulate
- Setting or social context of the interaction: where are they speaking? The manual is intended for internal communication
- The topic: what is being talked about? The subject is relevant to the agency and industry, in the first place, but also for ecologists and animal liberation activists, as source of information about what it is currently being done by the government and industry. It is a very specific topic with several ramifications that lead to a general public.
- The function: why are they speaking? The text is originated in a specific problem, i.e. there is a regulated social practice that hasn't been properly carried out. The manual, therefore, constitutes a kind of palliative measure that describes the right way in which the job must be done, using an informal register

Before indulging in the analysis to shed light on the institutional context in which the manual has been produced, several features pertinent to the text should be highlighted:

- The text has no direct authorship declared. It is presented as an institutional edition prepared by professional personnel of the agency. Five names are mentioned as responsible for the contents—not referred as authors—all of them hierarchical staff of the agency.
- The section of references and quotations in the body of the text has not been elaborated following the Vancouver Style (common in biological and veterinary sciences), Harvard, ACS, or APA Style formatting; instead, they are organized alphabetically by title and most data is missing: in several cases, place, publisher, and date are not present. In the case of web pages, access date is not specified
- The manual's date of publication is not mentioned, although the download link includes the year 2016 in the file name

This sloppy presentation of the official procedures' manual gives the text an appearance of a notebook, made roughly for an internal circulation. But it also reduces the intended appearance of scientific rigour making it look like a hastily written and poorly prepared text in the eyes of the public outside the agency.

Following Hodge and Kress (1993) in the discursive analysis, ideational meta-function in the preface shows three types of actors: (1) abstract concepts/nominalizations, (2) non-human beings, and (3) human beings. Relational, existential, and non-transactive processes. This means that participants are constructed as entities that evidence features and qualities. Remarkably, when they develop some kind of event, it has no impact or consequences on other participants.

Abstract Concepts

Animal welfare, welfare conditions, minimum stress situations, ethical aspects, animal production, among others. In some cases, the abstract concept refers to discrete, immaterial entity, like in the case of *ethical aspects*. But in others, as in *minimum stress conditions, animal productions*, or *welfare conditions*, they are implied in the context of particular agents, processes, and affected participants.

Abstract concepts are defined in terms of *representing, constituting*, and *projecting* other concepts, in the way of the dictionary. In terms of animal welfare, abstract concepts subsume human responsibility in constructions as in Table 2.1. Particularly, environmental conditions for farmed animals are imposed by their raisers and, therefore, animals face these conditions with the limited resources that the raisers allow. But this is not openly manifest in the text. These material processes contained in the abstract concepts are related—through relational processes—to other concepts that are "ethically friendly": *environmental conditions, ethical aspects, and compassionate treatment.*

Table 2.1 Abstract Concepts

El bienestar animal	Representa	el modo en que un animal afronta las condiciones de su entorno
Animal Welfare	*represents*	*the way in which an animal faces their environmental conditions*
El bienestar animal	Resulta	un aspecto ético
Animal Welfare	*constitutes*	*an ethical aspect*
El bienestar animal	se proyecta	en el trato compasivo
Animal Welfare	*is projected*	*in a compassionate treatment*

Non-humans

In the text, *animals* are only mentioned in relational or existential processes that link them to welfare conditions in positive ways, as we can see in Table 2.2.

Table 2.2 Relational and Existential Processes

Un animal	Experimentar	Las cinco libertades
An animal	*experience*	*the five liberties*
Un animal	Se encuentra	En condiciones de bienestar
An animal	*Is*	*in welfare conditions*

Humans

Humans are only mentioned indirectly in the last paragraph of the preface as the beneficiaries of the products obtained *through daily working routines and*

Table 2.3 Humans in Processes

El hombre	Obtiene beneficios	Animales domésticos
Men	*obtain benefits*	*domestic animals*
El hombre	Usa	Animales domésticos
Men	*use*	*domestic animals*

rentability indicators of the activity but also as consistent and responsible actors who react according to our times.

Two main frames are present in the text: scientific and neoclassical economics. They appear combined, suggesting continuity between one another. Processes that imply agency are transformed into nominalizations—abstract concepts—to avoid making explicit human responsibility. Therefore, *animal welfare represents the way in which an animal faces its environmental conditions*, which is another way to say *animal welfare is the group of measures we humans adopt to make farmed animals'* conditions of life not so terrible so as to not make them (the farmed animals) useless material.

In scientific discourse, actors are erased to give an idea of "non-biased discourse" but here, erasure also fulfills the function of making actors in violent actions disappear, as if the actions develop by themselves, without human participation. Neoclassical economics manifests openly in *use and production of domestic animals* for a human benefit. Following this logic, the five liberties that, according to the manual, an animal must enjoy considering they are living in good conditions, are just a quality checklist implemented for commercial purposes: domestic animals are going to be killed, but if they are relaxed in the previous hours to the slaughter, their meat is tenderer. On the contrary, in they spend their last hours in distress and tension, their meat becomes tough and lean, and therefore, less valuable, as the manual points out.

Spanish Text	**English Translation**
Un animal cansado posee menores reservas de glucógeno en los músculos, que posterior mente derivan en una incorrect amaduración de las carnes y la aparición de carnes os curas, firmes y secas o DFD en su sigla en inglés (2015, p. 114).	A tired animal has lower glycogen reserves in the muscles, which later lead to incorrect maturation of the meat and the appearance of dark, firm and dry meat or DFD in its acronym in English (SENASA, 2015, p. 114). [Author's translation]

Science and economics go hand in hand to justify the killing of animals for human consumption. Their mention as goods is explicit:

Spanish Text	English Translation
"El bienestar de los animales resulta una inversion que minimisa los problemas y las pérdidas en cadauna de las etapas y consecuentemente favorece la rentabilidad y productividad en cada esla bón de la cadena." (SENASA, 2015, p. XII).	"Animal welfare is an investment that minimizes problems and losses at each stage and consequently favors profitability and productivity at each link in the chain." (SENASA, 2015, p. XII). [Author's translation]

In addition, a small trace of green-washing discourse is found: at the end of the preface, the text manifests the aim to "propitiate the sustainable development of the activity." Although it constitutes a common practice in official documents in many countries, this mention in the context of an official animal welfare manual is a paradoxical element on the one hand, considering that it refers to an activity considered by international institutions (Food and Agriculture Organization of the United Nations, FAO, 2006, p. 84) as non-sustainable. On the other hand, it was not required by the topic of the text, suggesting political reasons for its mention. It can be argued that animal welfare does not represent a step towards sustainability in any case, although it has been claimed by the industry that it does (Verkuijl,, 2022). In any case, that discussion falls outside the limits of this research, but it is a topic also discussed within the limits of official discourses about animal welfare and environmental politics that calls for attention.

Image Analysis

The book contains seventeen illustrations—six photos and eleven diagrams. In this section we analyze the six photos and one diagram.

Systemic-functional analysis evidences, in the ideational meta-function, a symbolic-suggestive conceptual process that presents the horse in the close-up as the main participant. This kind of process represents meaning and identity as coming from inside the participant or derived from their qualities. In this case, the horse, in good shape and health, is feeding in the fields, giving a certain sense of stillness and tranquillity, deeply associated with nature and optimal conditions for life in freedom of a non-human animal. In this same sense, the background image of fields and other horses complements the concept of nature.

In the analysis of the interpersonal meta-function, it must be considered, as mentioned above, the close-up of the horse. This shot shows the horse as if the viewer is standing in front of him/her, in a "professional observer distance": close but not that close as to build a personal relationship. The horse is shown in an oblique angle, generating a feeling of detachment: he/she is not part of our world; we watch from the outside. Following the logic of scientific discourses, the image sets participants and viewers at the same vertical level, and relative power is built as equal, giving a sense of objectivity. Besides, the information in the picture does not interpellate the viewer directly; there is no participant gazing at the camera: it presents an act of offer. The image evidences a naturalistic modality, in the sense that it does not show saturation of color, electric effects, or science-fictional elements.

The textual meta-function can be analyzed in terms of composition: the horse is the main participant and is linked to the other horses in the back by their species features and with the fields by the idea of nature constructed in the picture. The given/new system evidences the horse as the new, the identifiable participant, topic of the book, and the given, the fields in the background, the context of this topic.

Figure 2.1 is present in the section called Marking, in page 11. In the ideational meta-function, we observe a symbolic-attributive-conceptual process. This kind of process represents meaning as being conferred to the participant—in

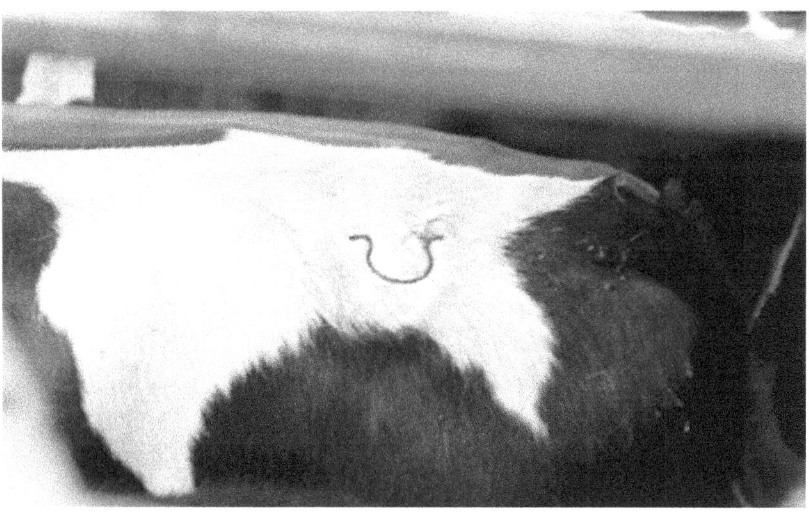

Figure 2.1 First illustration for the Section "*Marking.*" Source: National Service of Agri-Food Health and Quality (SENASA).

Figure 2.2 Second illustration for the Section "*Marking*." Source: National Service of Agri-Food Health and Quality (SENASA).

this case, by an actor that is not present in body and shape but whose effects can be seen in the mark. This conceptual process subsumes a material narrative process—the material action of marking that provoked the mark—and the human participants involved. The affected participant is the only one present and, in turn, he/she becomes participant in the symbolic attributive conceptual process.

Interpersonal meta-function shows an act of offer, there is no gazing at the camera. The very close-up shot focuses on the detail of the burning mark, and the blurring railing in the back completes the idea of naturalistic modality. The burning mark is taken directly in a frontal shot, constructing it as part of our world; another feature of the objectivity sought by scientific discourse. Nevertheless, in this case, power is on the viewer's side: the mark is taken from above, as a part of the scientific discourse.

The textual meta-function places the mark in the center of the image: it is the main significant element. The other, the blurry one, is the railing that reminds the viewer that they are watching a cow in a barnyard. The cow and the barn constitute the given elements and the mark is the new, the focus of the meaning. Figure 2.2 is also in the section "Marking," in page 11. This image somehow completes the meaning of the previous one in terms of "marking."

As in Figure 2.1, in the meta-function, a symbolic-suggestive conceptual process can be observed; in other words, meaning comes from the participant.

Figure 2.3 General principles about the facilities. Source: National Service of Agri-Food Health and Quality (SENASA).

The faces of the seven cows suggest attentiveness and fear. They are all staring at the camera, and their ears are all tense and backwards in order to get the best capture possible of any sound (Gleerup, 2017). The three cows from right to left are watching closely, but the one in the upper far left seems to have its eyes wide open, as if expressing fear. Above all, they are not relaxed.

The main element in the interpersonal meta-function is the act of demand posed by the cow's gazing at the camera. In the first place, it must be noted that they create a demand; they interpellate the viewer through gazing. But, on the other hand, the scientific discourse constructed in the manual presents them as commodities, and in the image, they are shown in group, with no particular differences with one-another. Therefore, the demand effect posed by the gazing is lost when the image is placed in the context of the economic-scientific discourse. Not even the close-up can reverse the idea that they are just commercial material. The vertical angle that constructs equal power on both sides of the camera reinforces the traces of scientific discourse.

In the textual meta-function, it is relevant to consider the given/new system; that is, the cows are the given while their faces represent the new: what they feel. Figure 2.3 illustrates the section "General principles about facilities" in page 97.

This picture presents an arm used to move animals to different areas and transports. Its importance is exclusively related to the management of the animals. Therefore, ideational meaning is constructed by absence: this place is

Figure 2.4 Particular aspects of animal handling during events. Source: National Service of Agri-Food Health and Quality (SENASA).

for the animals, but they are not present. It is a symbolic attributive conceptual process that needs to be completed with the absent participants. As viewers, we know what happens in there, especially since animals are required for the process to take place. Consequently, its significance is related to the absent participants.

Interpersonally, the image is information offered, an act of offer in a medium shot. It sets us, as viewers, close enough to see but not close enough to be a part of it. The angle of the take is oblique. This puts us away: this is not our world; we are looking from the outside. Textually, the entrance is located at the center of the image, placing it as the main element in the picture. Figure 2.4, illustrates, in page 99, the section called *Particular aspects of animal handling during work day or event.*

The analysis of the ideational meta-function shows a symbolic-attributive conceptual process: the animals have been put inside the barns for sale, they are material for sale. In this sense, they are also affected participants in a material process with a human agent missing.

Interpersonally, camera angle places relative power on the side of the viewer. Cows are taken from above, looking inferior to the viewer. The image presents information offered; there is no act of demand. Naturalistic modality: animals are not individualized but they are all parts of the group, they are shown as uncountable material, not individuals.

The textual meta-function shows the cows as the given element and the barn fence as the new: the given-new system tells us that we had been talking about

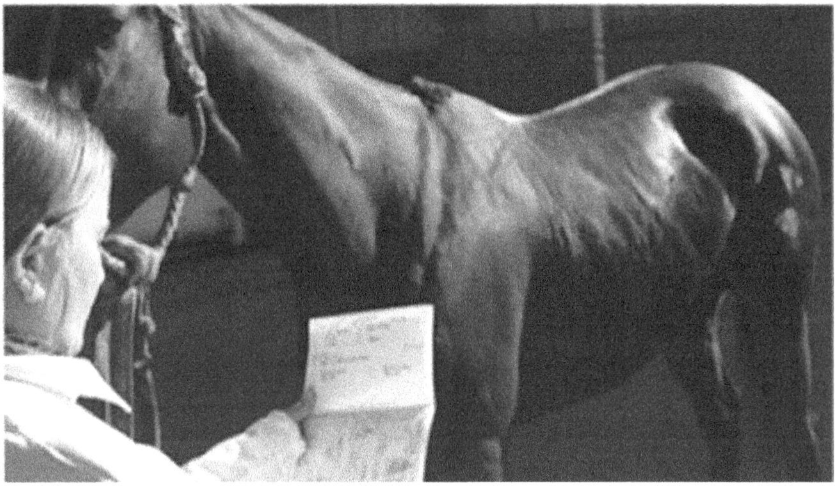

Figure 2.5 Illustration for the Section "*Control*." Source: National Service of Agri-Food Health and Quality (SENASA).

cows, now we are talking about cows for sale in a barn. Figure 2.5 belongs to the same section as Figure 2.4, in page 42, and exemplifies documentation controls.

In this case, ideational meta-function shows two narrative transactive processes: a human participant is controlling the horse's documents while the other, whose hand is only visible, is holding the horse by the rein. In both processes, the horse is the affected participant. The white apron of the woman controlling the documents suggests she is a veterinarian or a SENASA agent. In any case, her look points in the direction of scientific discourse.

The interpersonal metafunction builds closeness through a close-up of the woman and the documents, in an act of offer that also adds to scientific discourse. In the textual meta-function, humans are shown as the given and the horse as the new, which has to be verified. In this picture, human position relates to the human position in the text of the manual: humans manage non-humans.

The manual contains fifteen diagrams that can be classified into two groups: those *showing* conceptual analytical topological processes and those *containing* conceptual analytical topographic processes. In both cases, diagrams add a technical-scientific value to the visual discourse, explaining in detail how animal behaviors can be considered or how to perform a certain practice on the animals.

In page 21, under the title "*natural behaviour and species-specific aspects*," we find this conceptual analytical topographic process (Figure 2.6). Elements

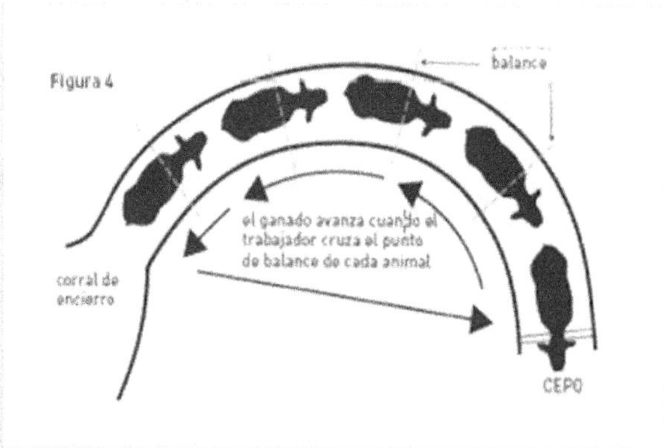

Figure 2.6 Conceptual analytical topographic processes.Source: National Service of Agri-Food Health and Quality (SENASA).

represent precisely physic-spatial coordinates, relations, and locations. On the contrary, topological processes represent logic relations in detail.

In this case, we can see an explanation of how cows must walk through the arm and where workers must place themselves during the operation. Human participants, actors with agency in the represented process, appear only in the word "workers"; there is no drawing to represent them. On the other hand, cows are well represented in the image.

Conclusions

Verbally and visually, the scientific-neoclassical economic articulation constructs a narrative describing the relationship between humans and non-humans. In the prologue, the terms of this relationship are openly described, and this points out that the manual has been elaborated for the development of the industry. Images help building a discourse in which science is put at the service of economic growth following the logic of naturalistic modality: non-humans are part of conceptual processes that show them *being* and not *doing*. Humans, on the other hand, are the actors behind processes that affect non-humans. Non-humans are constructed through pictures in an optimal state and not in the real conditions they face during the food production chain. Damage provoked by the lack of animal welfare is mentioned but not evidenced in pictures. In those cases, naturalistic modality is lost and diagrams illustrate situations in which trauma can hypothetically appear. In contexts where damages can be seen, pictures show long shots that do not allow seeing particular features such as injuries.

Within the manual, the Argentine agrifood system is constructed in a way that not only ignores cruelty to non-humans but also promotes economic growth through unsustainable activity (FAO, 2006, p. 87). It also favors Argentina's landowners, the ruling class since the foundation of the country in 1810 (Sáenz Quesada, 2013, p. 141). In this context, official discourses have the opportunity to change the way we talk about non-humans, the climate crisis, and social conflicts; the objective is to take steps towards real change by enforcing social practices that transform the way the ruling classes think about nature and environment. In democratic societies, the citizens have the power to make governments enforce this.

References

Coulthard, M., Johnson, A., & Wright, D. (2017). *An introduction to forensic linguistics. Language in evidence* (2nd ed.). Routledge.

European Commission. (2020). *Farm to Fork strategy for a fair, healthy and environmentally-friendly food system.* Retrieved from: https://food.ec.europa.eu/horizontal-topics/farm-fork-strategy_en

Flax, R. (2020). El análisis del discurso como herramienta para la enseñanzade epistemología [Discourse analysis as tool for teaching epistemology]

Food and Agriculture Organization of the United Nations (FAO). (2006). *Livestock's long shadow: Environmental issues and options.* FAO.

Forte, D. (2019). La construcción del discurso identitario argentino : la historia de la carne [The construction of the discursive identity of Argentina: The story of meat]. *Question, 1*(64). Universidad de La Plata.

Foucault, M. (1978). *La verdad y las formas jurídicas* [Truth and legal forms]. Editorial Gedisa.

Gleerup, K. (2017). Identifying pain behaviors in dairy cattle. *WCDS Advances in Dairy Technology, 29*, 231–239.

Hodge, R., & Kress, G. (1993). *Language as ideology.* Routledge.

Holmes, J. (2008). *An introduction to sociolinguistics* (3rd ed.). Person Education

Kress, G., & van Leeuwen, T. (2005). *Reading images. A grammar of visual design.* Routledge.

Lakoff, G. (2002). *Moral politics: How liberals and conservatives think.* University of Chicago Press.

Raiter, A. (2003). *Lenguaje y sentido común: las bases para la formación del discurso dominante* [Language and common sense: Foundations of the dominant discourse]. Biblos.

Sáenz Quesada, M. (2013). *Los estancieros: Des de la época colonial hasta nuestros días* [Ranchers. From colonial times to our days]. Sud Americana.

Servicio Nacional de Sanidad y Calidad Agroalimentaria (SENASA). (2015). *Manual de bien estar animal. Unenfoque práctico para el buen manejo de especies domésticas, durante sutenencia, producción, concentración, transporte y faena* [Animal welfare manual. A practical approach to the correct management of domestic species, during their possession, production, concentration, transport and slaughter]. SENASA. Retrieved from https://www.argentina.gob.ar/senasa/bienestar-animal/manuales

Stibbe, A. (2015). *Ecolinguistics: language, ecology and the stories we live by*. Routledge.

Verkuijl, C., & Sebo, J. (2022, May). *Mainstreaming animal welfare in sustainable development: Stockholm+50 background paper*. Stockholm Environment Institute. https://www.sei.org/wp-content/uploads/2022/05/animal-welfare-stockholm50backgroundpaper.pdf

Verón, E. (2011 [1987]). *La semiosis social: Fragmentos de una teoría de la discursividad* [Social semiosis. Fragments of a discursivity theory]. Gedisa.

Vilajosana, J. M. (2010). La construcción social de la realidad jurídica [The social construction of legal reality]. *Prisma Social*, *4*, 1–19.

3

Grammatical Reflections of Sentient Landscapes

Linguistic Evidence for Environmental Personhood

David Stringer

Environmental personhood—the granting of status as persons before the law to natural entities such as rivers, mountains, forests, or the whole of nature, as a means of protection from despoilment—has emerged as a powerful legal concept in recent years. Since the approval of a new Ecuadorian constitution recognizing the rights of *Pacha Mama* "Mother Earth" in 2008, cases advocating for environmental personhood have been pursued in over forty countries, although the concept has been understood in several distinct ways and is often misconstrued. This paper explores the origins of non-human personhood in European and American law and identifies two different approaches to legal guardianship, one involving a metaphorical understanding of the concept, akin to corporate personhood, and the other resting on a literal belief in non-human right to life. In cases resting on the idea of literal personhood, success in court may depend, in part, on whether those proposed for guardianship are seen, on the one hand, as opportunistic conservationists propagating revisionist religion, or, on the other hand, as authentic Traditional Owners of lands and waters. For many Indigenous communities, the prospects for survival of ancestral language, customs, and environment are profoundly linked to efforts to secure land tenure. Culture is rooted in ecology. Moreover, inhabited landscapes are not just spaces that encompass human activity, but expanses of life, with plants, fungi, and other animals that many people understand to be enspirited. Therefore, the struggle to protect ancestral lands is often part of a broader effort to establish more-than-human rights, commensurate with worldviews that do not recognize a clear dividing line between culture and nature. While arguments can be made on the basis of cultural or religious beliefs, more concrete evidence for the deep cultural grounding of non-human persons can involve the encoding of features such as animacy and personhood in particular language grammars. A preliminary crosslinguistic survey suggests that these features are manifested in all languages but are applied in culture-specific ways. As such, communities seeking rights for sentient landscapes have the potential to invoke legal evidence for established beliefs in the form of grammatical reflections of environmental personhood.

From Corporate to Environmental Personhood

It is important at the outset to distinguish between human and legal personhood, as this has been a major point of confusion both in the news media and in academia beyond legal scholarship, in the humanities and social sciences. Throughout the history of European and American law, legal personhood has traditionally been extended to entities that no one would consider animate entities such as trusts, corporations, churches, and states. This has long been considered a legal necessity, so that, for example, the rights, responsibilities, and penalties of an organization do not evaporate when a human leader who signs a document dies. Corporate personhood has its origins in Roman law, and was developed in medieval Europe as scholars came to terms with rights to inheritance, rights to sue, and peace treaties concerning organizations that went beyond the individuals representing them (Avi-Yonah, 2005; Pollman, 2012). If a King or a Pope signed an agreement, they clearly staked their own honor on such a contract, but what was to happen to the relevant rights and obligations of the State or the Church after the death of the signatories? The need for organizations to be treated as persons under the law was extended throughout society, such that a business was also understood to be able to sue or be sued, and to own property. More recent precedents of modern legal arguments of corporate personhood can be found in nineteenth and twentieth-century American law. In early debates, the main concern was often to provide legal mechanisms to protect the status of religious and educational institutions; it was understood that the people managing schools and churches should not be considered "owners," whose properties might be inherited by their children (Blair, 2003). Thus the property of an institution came to be owned by the institution itself, with humans simply managing it, for the benefit of present and future generations. The possible implications of this for environmental personhood should be clear.

The legal concept of corporate personhood nevertheless mutated considerably during this time, and its implementation continues to be contentious to the present day. Over the years, corporate personhood has been operationalized in terms of Concession Theory, Contract Theory, and Real Entity Theory (for an accessible overview, see Gordon, 2019). The basic understanding of Concession Theory was that that personhood status was granted by concession of the state. This was the approach that characterized early corporate charters in the United States, when such cases were relatively rare (Avi-Yonah, 2005). Various issues related to the power of the state to grant or deny status led to the development of Contract Theory, which rested on the concept that organizations consist of groups of human beings and form themselves through contractual agreement, rather than according to the whims of a particular government. This, however, led to problems of limited liability (Harris, 2006, p. 1470). Current approaches to corporate personhood in the United States are framed in terms of Real Entity Theory, according to which the corporation is an actual legal person, with constitutional protections based on the Bill of Rights and the Fourteenth Amendment, which states that all individual citizens have equal protection under the law (Harris, 2006, p. 1473).

Following extremely controversial decisions by the US Supreme Court that have relied on the concept of corporate personhood, there has been widespread outrage at the idea that corporations can be treated as equivalent to people. In Citizen's United (2010),[1] the court overturned restrictions on corporate interference in elections, allowing unlimited speech and spending by corporations (and fueling the rise of Super PACs). This decision was and remains very unpopular (ABC-Washington Post poll, 2010: 80 percent opposed, 65 percent strongly opposed). In the Hobby Lobby (2014) case,[2] corporations were granted to right to deny their employees health care coverage for contraception, which was met with similar disbelief (the American Congress of Obstetricians and Gynecologists, whose membership includes 90 percent of gynecologists, backed a bill to reverse the ruling; Bassett, 2014). However, while such decisions may indeed be egregiously misguided for a host of reasons, outrage is often misdirected.

Arguments that corporate personhood is ridiculous—how can a corporation be a person?—are often made in ignorance of the fact that the notion is has been generally accepted for centuries in legal circles, where legal personhood is rarely confused with being a human person. To articulate the absurdity of the claim that corporations are actually people is in many ways to attack a straw man. This disconnection was apparent in the controversy over the politician Mitt Romney's comments at the Iowa State Fair in 2011, where he answered a heckler by saying "Corporations are people too, my friend," resulting in years of mockery in the press. This blurring of human and legal personhood is also to the fore in Stibbe's (2015, pp. 75–81) influential criticism of the metaphor THE CORPORATION IS A PERSON , although it could be argued that this is a fair analysis given the extreme interpretations of Real Entity Theory by a rogue five-person majority on the Supreme Court (see the coruscating ninety-page dissent in Citizens United by Justice John Paul Stevens, joined by the other three justices).[3] Anger at the arguable abuses of corporate personhood may need to be appropriately contextualized as this legal strategy is extended to the environment, such that a landscape or a river may conceptualized as an entity that owns itself, managed by and legally represented by specific groups of people who serve as temporary caretakers of places that will still be there long after the current generation of guardians is gone.

Just as with corporate personhood, the news media, politicians, and certain corners of academia often ridicule the notion of environmental personhood: how can a river be a person? Following the first US Supreme Court consideration of an (unsuccessful) environmental personhood case, the dissent by Justice William O. Douglas foresaw that contemporary environmental concern "should lead to the conferral of standing upon environmental objects to sue for their own preservation."[4] In response, he was lampooned in the American Bar Association Journal: "If Justice Douglas has his way—/ O Come not that dreadful day—/ We'll be sued by lakes and hills /Seeking a redress of ills" (Naff, 1972). More recently, following the failed attempt to gain recognition of the Colorado River ecosystem as a legal person, Senator Steve Daines from Montana opined, "Radical obstructionists who contort common sense with this sort of nonsense undercut credible conservationists" (Turkewitz, 2017). These kinds of uninformed pronouncements by benighted opinionators are of a kind with failed reductions to absurdity regarding corporate personhood, as discussed above.

However, there is a twist to the story as concerns environmental personhood. Unlike corporations, environments are actually alive, so there is, in fact, an entirely different kind of legal argument to make. One of the most influential sources for contemporary discussion of environmental personhood is Stone's (1972) revolutionary and prescient paper *Should Trees Have Standing?* In this forward-looking think-piece, Stone first acknowledges that most readers might find absurd the idea of extending rights to "natural objects" such as forests, oceans, and rivers. He then assembles an array of observations and propositions to make the unthinkable thinkable. At various points in Western legal history, it would have appeared absurd to makers of the law to grant full juristic personhood rights to all people, including children, women, or people of color. From today's perspective, we ask how it could ever have been legal for a father in Ancient Rome to kill or sell one of his own children, for the Supreme Court of Wisconsin to deny a woman the right to practice law because such a thing would be a "[departure] from the order of nature,"[5] or for the Supreme Court of Virgina to declare, regarding African Americans, that "so far certainly as civil rights and relations are concerned, the slave is not a person, but a thing"?[6] In our own times, the denial of such rights seems universally unconscionable, which should make us wary of what today might seem to us obvious limitations on the nature of literal personhood. Stone argues that there is no logical reason why we should not extend rights to non-human forms of life (e.g., fish, trees), which can also be thought of as having intrinsic rights, and the environments that support them (e.g., rivers, forests), in terms of an understanding of such natural entities has having a right to life. He clarifies that this does not mean that no one should be allowed to cut down a tree; after all, several countries allow a person to be executed; it is a case of balancing rights, rather than one party possessing and another being devoid of rights (p. 457).

While the ground-changing potential of Stone's (1972) argumentation would be quickly picked up and broadcast widely in the Sierra Club vs. Morton dissent mentioned above, it would be many years before legal recognition of the rights of nature became a reality. One important factor in the spread of what was fundamentally an environmentalist polemic was that it relates harmoniously to much older ideas. To many Indigenous people, social and psychological personhood extends beyond the human; ancestral lands, waters, and the animals and plants they support are enspirited: they have souls, although this term resists easy translation. The ground had been laid for possible joint efforts by conservationists and Indigenous activists to embark on a new kind of legal strategy to protect ancestral lands from environmental degradation, and over the last two decades, cases involving environmental personhood have proliferated throughout the world.

The Rise of Environmental Personhood as a Legal Strategy

One relatively small case in the United States was arguably the catalyst for the prominence of environmental personhood as a legal strategy today. In

2006, a nonprofit group called the Community Environmental Legal Defense Fund (CELDF) worked with representatives of the community of Tamaqua, in Schuylkill County, Pennsylvania, to craft a law that would prevent sewage haulers from spreading toxic sludge on farmland with the blessing of the farmers and the backing of the authorities. The new ordinance allowed not only the municipal government but any local resident to file a lawsuit on behalf of the environment, completely upending the legal precedent of treating nature only in terms of property. For the first time anywhere, Nature was officially understood in written law as holding rights in and of itself. Not only has the same organization gone on to draft similar legislation for nearly 200 municipalities across the United States (CELDF, 2024), but it served as inspiration for revolutionary constitutional changes at the national level in Ecuador.

Largely as a consequence of Indigenous voices claiming greater political power, Ecuador held a national referendum in 2008, after which it changed its constitution to reflect rights for *Pacha Mama*, a female spirit/goddess whose name is sometimes translated as Mother Earth or Mother Nature, but who resists simple personification. Any individual or organization can advocate on behalf of Nature in general, which now has constitutional rights. Article 71 states: "Nature, or Pacha Mama, where life is reproduced and occurs, has the right to integral respect for its existence and for the maintenance and regeneration of its life cycles, structure, functions and evolutionary processes," and Article 73 further clarifies, "The State shall apply preventive and restrictive measures on activities that might lead to the extinction of species, the destruction of ecosystems and the permanent alteration of natural cycles. The introduction of organisms and organic and inorganic material that might definitively alter the nation's genetic assets is forbidden" (Ecuador National Assembly, 2008). Bolivia, which had also seen a transformation in Indigenous political representation, soon followed suit, with the difference that new constitutional rights were granted in 2010 for the benefit of a "collective public interest" (Plurinational State of Bolivia, 2008, Ch.II, Art. 5); that is, nature is not enshrined as a legal person apart from humans, but as a system including humans, in which humans rights do not predominate with respect to the rights of other aspects of the natural world. Both countries were leaders in raising consciousness of this fresh approach to environmental conservation at the United Nations, and spurred action in other countries.

While Ecuador and Bolivia granted rights to nature as a whole, other countries have opted to developed more targeted legislation, aimed at protecting particular landscapes or waterways. In 2014, Aotearoa New Zealand became the first country to cede ownership of a national park when Te Urewera became a juristic entity with "all the rights, powers, duties and liabilities of a legal person" (TNZPCC, 2014, Clause 14, 1). For Māori communities, a forest, mountain or a river might be a *tupuna* ""ancestor" for an *iwi* "tribe," *hapu* "sub-tribe," or *whanau* "extended family group," and this *whakapapa* "genealogy" is integral to the realization of the law. The local Te Urewera Board, comprising both Tuhoe people for whom this is ancestral land and representatives of the Crown, can bring lawsuits on behalf of the land itself. Soon afterward in 2016, legal personhood was granted to Te Awa Tupuna (the Whanganui River), after what had amounted to 170 years of litigation by Māori people. Again personhood is literal, although in this case there are two guardians, Te Pou Tupana, chosen by iwi and the Crown, who act on behalf of Te Awa Tupuna and confer with the local administrators, iwi, and hapu of the river.

Some of the other more than forty cases successfully advocating for environmental personhood worldwide since Ecuador's landmark change to its constitution include the granting of inalienable rights to the Atrato River in Colombia by the Constitutional Court of Colombia (2016); having closely followed the development of the Te Awa Tupuna case, the court mandated a guardianship committee for the river, initially with one but now fourteen council members. A 2017 ruling by Indian's High Court of Uttarakhand granted legal personhood to the Ganga River Basin, and another ruling shortly afterward by the same court extending personhood to "glaciers including Gangotri and Yamunotri, rivers, streams, rivulets, lakes, air, meadows, dales, jungles, forests, wetlands, grasslands, springs, and waterfalls" (Miglani vs. State of Uttarakand, 2017, Clause 2). While many cases in the United States and Canada have met with resistance, often facing years of appeals and counterlegislation, there have been some notable recent successes and new initiatives: in 2021, Muteshekau-shipu (The Magpie River) in Quebec became the first in Canada to be granted juristic personhood, by the local municipality of Minganie and the Innu Council of Ekuanitshit; in 2018, the White Earth Nation, Minnesota, passed a law recognizing manoomin, a sacred wild rice, as having inherent legal rights, and in 2021 they sued the Minnesota Department of Natural Resources on behalf of the manoomin, as a proposed gas pipeline violated the rice's right to "exist and flourish"; in 2022, the Sauk-Suiattle tribe sued the City of Seattle on behalf of the *TsuladxW* "salmon," arguing for dam removal.

Of course, success cannot be measured at the moment of a ruling, but on what happens subsequently both in the courts, on the ground, and in the waters.

Environmental destruction continues apace in Bolivia (Paredes Tamayo, 2024), the Ganges is still polluted (Das, 2021), and the White Earth Nation case was dismissed (Sleeper, 2023). Other cases have seen more optimistic outcomes, whether in terms of compromise or outright victory. The City of Seattle agreed to provide safe passage for TsuladxW around three hydroelectric dams; and in Ecuador, people voted in a 2023 referendum to halt all oil drilling in the Yasuní territory in the Amazon, a place with rich Indigenous culture and extremely high biodiversity. As legal argumentation continues to proliferate, and as it becomes more apparent that current cases depend more on literal than metaphorical personhood, it is important to guard against the kind of attacks invoking absurdity that have been leveled against non-human personhood for the last two centuries. Two related types of reasoning seem particularly important: that many Indigenous communities do not draw the line between persons and nonpersons at a juncture between humans and other species; and there is often solid evidence that such beliefs are longstanding because they may be encoded in the grammars of Indigenous languages.

Decolonizing Assumed Dichotomies

It should now be clear that for many of the communities involved in recent and current litigation, environmental personhood is literal, not metaphorical. Sentient landscapes are more than legalistic concepts. While many people in Western traditions of thought see forests, rivers, mammals, birds, fish, and plants as *things*, others all over the world (including in Europe and North America), understand such entities as *beings*, with animacy. The dichotomy between animate and inanimate is also related to a distinction between entities with and without souls, and ultimately concepts of personhood.

Even well-meaning anthropologists have struggled to escape misleading characterizations of ecosystemic relations due to their inherited assumptions about distinctions between humans and other animals, between animals and plants, or between culture and nature. For example, Reichel-Dolmatoff (1997) attempts to paint a positive portrait of Tukano shamans in the Colombian rainforest as Indigenous ecologists. He describes their relationship with the forest in terms of the "control and management of natural resources," and he argues that religious practices can be understood in terms of "one single problem—the maintenance of a balanced ecosystem" (pp. 7, 15). In his dealings with the spirits, the shaman can be thought of as an "ecological broker" whose "book-keeping shows the general system inputs and outputs" (p. 18). Such a metaphor depends on acceptance of a division between humans and other animals, and between society and nature. However, the analogy invoked in this case is highly questionable, because, as Descola (2013) notes, the Tukano ascribe to both animals and plants the psychological and social attributes of humans (pp. 12–13). In their dealings with the natural world, people negotiate with animals and plants as social partners, and do not see them as units of accounting.

In fact, it is typical of rainforest peoples with animist belief systems to see nature as lacking clear psychological and social boundaries with humans. In Ecuador, Achuar hunters understand toucans and howler monkeys to be brothers-in-law, subject to similar rules concerning killing in-laws in the course of a vendetta. Achuar women talk to the plants in their garden using a language of address otherwise reserved for children. The souls of animals and plants are social partners in an environment not distinct from human society (Descola, 2013, pp. 4–6). In New Guinea, for Karam people, the killing of *kobtiy* "cassowaries" involves highly prescribed behavior, as these birds are traditionally considered to be female cross-cousins of human origin (Bulmer, 1967). In many cultures, not only animals but certain trees are understood to have human-like souls. Dogon healers in Mali consult trees to ask them about their knowledge of the forest, and some trees, such as the kapok, are believed to move around at night, having conversations with other trees in the forest (Descola, 2013, p. 27).

People have considerable choice in how they conceptualize the outside world, and whether they perceive a separation between humans and other animals, between animals and plants, or between culture and nature. Lest these examples seem exotic to some Western readers, it is perhaps worth remembering that animist beliefs in animal and plant souls are not so alien to dominant world cultures. As discussed by Trumper (2004, p. 69), the Ancient Greeks believed that plants had a vegetative soul, animals had a sentient soul, and humans had a thinking, or rational soul. In almost perfect parallelism, early Chinese philosophy posited that plant souls consisted of spirit + life, animal souls contained spirit + life + perception, while human souls were composed of spirit + life + perception + justice/morality, concepts reflected in the different orthographic characters for "soul." The basic dichotomies of person vs. thing, animate vs. inanimate, or or culture vs. nature, so often assumed in law, are clearly subject to cultural relativity.

Linguistic Reflections of Sentient Landscapes

One way to disarm those arguing against literal understandings of environmental personhood from point of view that such claims are just "nonsense" (see discussion of Senator Steve Daines above) is to demonstrate that a belief in the animacy of landscapes or rivers is not just revisionist religion or juristic opportunism, because such cultural understandings have often been encoded in grammar over hundreds or thousands of years. To my knowledge, the linguistic encoding of animacy in Indigenous languages has not been explicitly discussed or cited as evidence in cases of environmental personhood to date. However, this has tremendous potential to demonstrate to outsiders the longstanding beliefs held in many communities whose environments are currently under threat.

A survey of typological research in linguistics reveals that personhood is universally grammaticalized in the world's languages, though with different forms of exponence (see e.g., Comrie, 1989, pp. 185–200; Croft, 2003, pp. 128–142; Santazilia, 2022). There appears to be an implicational hierarchy i that indicates possible cut-off points for personhood in any given language. Croft (2003) argues that it involves three related functional dimensions: Person: third > second, first; Referentiality: common noun > proper name > pronoun; Animacy: inanimate > animate > human. Linguistic realization involves interaction with other concepts, such as definiteness and specificity. Some examples of grammatical exponence are as follows.

Expression of personhood is quite typically found in pronominal systems. In English, while first and second person pronouns do not encode personhood (they are used in dialogue and are presumed animate), third person pronouns are split.

(1) See {that girl / that forest}? {She's / It's} wonderful.

This is also true for relative pronouns (*who* vs. *that*).

(2) *thc book who I read last week

In Te Reo Māori (the Māori language), the *wh*-phrase is different for enspirited landscapes and inanimate objects, revealing grammatical encoding of environmental personhood (Te Ngaru Weki, personal communication, May 13, 2024).

(3) a. Ko hea tō maunga? b. He aha tō kai?
Who-is your mountain what-is your food
"Who is your mountain? "What is your food?

Another kind of grammatical exponence of personhood is found in case-marking of direct objects. For example, in Hindi, *–ko* is used to animate direct objects. Actual use is somewhat complex: it is required with specific animate NPs; optional with specific inanimate NPs; and usually ungrammatical with any non-specific NP (Montrul, Bhatt, & Girju, 2015, p. 570).

(4) Mira-ne ramesh-ko dekhaa. (*Mira-ne ramesh dekhaa.)

Mira-ERG Ramesh-DOM saw
"Mira saw Ramesh.'

Similarly, Spanish marks personhood specific direct objects. Thus in northern Peru, where sentient landscapes are seen as agents of social justice, and mountain spirits have been named as plaintiffs in environmental cases against extractive industries (Bacigalupo, 2021), when a *wak'a* (landscape made by indigenous ancestors, e.g., pyramid, tomb) or *apu* (ancestor mountain) is used as a direct object, they take the "personal *a*." Moreover, such entities are not understood as places that houses spirits, but rather as beings themselves, with proper names (Mariella Bacigalupo, personal communication, Feb 20, 2024).

(5) Vamos a ver {a la Wak'a Cao / al Apu Cuculicote}.

Go-1PL to see {DOM the Wak'a Cao / DOM-the Apu Cuculicote}
"Let's go and see {DOM the Wak'a Cao /DOM-the Apu Cuculicote}.'

Types of grammatical expression indicative of personhood take a wide range of forms across languages, and any given language may have multiple and sometimes surprising ways to show linguistically whether an entity is a person or not. To return to English, for example, prepositional passives are licensed only by verbs that select for animacy.

(6) a. This bed was slept in by my grandmother.
b. *The horizon was appeared on by the sun.

Chinese expresses personhood in numerical classifiers (counting persons with *ge* is different from counting books or pens); Japanese does do in verbs of existence (*iru* "to be" is for living beings and *aru* "to be" is for inanimate entities). Even basic word order may depend on an animacy hierarchy, which can sometimes interact with first and second vs. third person. For example, in Diné (Navajo), basic word order depends on the position of entities in the great chain of being, e.g., human > bear > cat > ant > wind > tree > abstractions. Thus one cannot say something akin to *The cat bit the bear*, but rather something roughly rendered as *The bear the cat bit*, with relevant marking on the verb to indicate who is doing the biting (Evans, 2022, p. 65). In the language Southern Tiwa, these kinds of considerations apply also to pronouns, with first and second person ranked higher than third; therefore, the order *The man saw me* is impossible, and must be expressed by something analogous to *I was seen by the man* (Comrie, 1989, pp. 192–193).

In sum, when Indigenous communities seek rights for sentient landscapes and waterways, and must explain to outsiders how exactly they conceive of enspirited natural phenomena, one convincing form of evidence for time-honored beliefs is the grammatical encoding of personhood. Not all languages will encode personhood in the relevant way for the particular entities under consideration, but it certainly behooves lawyers to take a look.

Conclusion

The interests of conservationists and Indigenous rights activists are not identical but are often aligned in terms of land tenure and environmentalism. While argumentation for environmental personhood may seem to flow naturally from the long-established tradition of corporate personhood, with human guardianship established to take care of an entity before handing it on to future generations, the reality is that modern litigation is more often in terms of literal rather than metaphorical personhood. In conservation circles, this line of thought flows from Stone's (1972) pioneering polemic, but in actuality, many Indigenous communities around the world have understood such concepts since time immemorial. Following a half-century of judges and politicians raising the question of whether assigning personhood to natural entities even merits serious consideration, it is time to add linguistic evidence to the mix. Depending on the language, a mountain may be a *she* not an *it*, a *who* not a *what*, be marked for personhood when the recipient of an action, or precede a hut in every sentence that puts them together. For Indigenous cultures seeking environmental protection, such linguistic encoding is a dependable source of evidence for deep-seated beliefs in sentient landscapes.

> *So simply to look on anything, such as a mountain, with the love that penetrates to its essence, is to widen the domain of being in the vastness of non-being.*
>
> Nan Shepherd, *The Living Mountain*

Acknowledgments

These ideas benefitted significantly from brainstorming and joint discussion with students Ava Hartman and Kamebry Wagner at Indiana University Bloomington. Also, much love and thanks to Pricilla and Te Ngaru Wetu, and to Mariella Bacigalupo. Any remaining errors are my own.

Notes

1 Citizens United v. Federal Election Commission, 558 US 310 (2010).
2 Burwell v. Hobby Lobby, 573 US 682 (2014).
3 Citizens United v. Fed. Election Comm'n, 558 US 310 (2010) (Stevens, J.P., dissenting).
4 Sierra Club v. Morton, 405 US 727 (1972) (Douglas, W.O., dissenting).
5 In the Matter of the Motion to admit Miss Lavinia Goodell to the Bar of this Court, thirty-nine Wis. 232(1875), at 244–245.
6 Bailey v. Poindexter's Ex'r, fifty-six Va. (14 Gratt.) 132, 14,2–43 (1858)].

References

Avi-Yonah, R. -S. (2005). The cyclical transformations of the corporate form: A historical perspective on corporate social responsibility. *Delaware Journal of Corporate Law*, *30*(3), 767–818.

Bacigalupo, A. M. (2021). Subversive cosmopolitics in the Anthropocene: On sentient landscapes and the ethical imperative in Northern Peru. In E. Berry (Ed.), *Climate politics and the power of religion* (pp. 176–205). Indiana University Press.

Bassett, L. (2014, July 14). Major doctors' group endorses Hobby Lobby override bill. *The Huffington Post*. https://www.huffpost.com/entry/major-doctors-group-endor_n_5585018

Blair, M. M. (2003). Locking in capital: What corporate law achieved for business organizers in the nineteenth century. *51 UCLA Law Review, 51*(2), 387.

Bulmer, R. (1967). Why is the cassowary not a bird? A problem of zoological taxonomy among the Kalam of the New Guinea Highlands. *Man, 2*, 5–25.

CELDF. (2024). About CELDF. https://celdf.org/about-celdf/

Constitutional Court of Colombia-. (2016). The Atrato River Case (English translation). http://files.harmonywithnatureun.org/uploads/upload838.pdf

Comrie, B. (1989). *Language Universals and Linguistic Typology* (2nd ed.). Blackwell.

Das, B. (2021). Toxic plastics choking the River Ganges. *Nature India.* https://www.nature.com/articles/d44151-021-00027-9

Descola, P. (2013). *Beyond nature and culture.* (J. Lloyd, Trans.). University of Chicago Press.

Ecuador National Assembly (2008). Constitution of the Republic of Ecuador (English translation). https://pdba.georgetown.edu/Constitutions/Ecuador/english08.html

Evans, N. (2022). *Words of wonder: Endangered languages and what they tell us.* Wiley.

Gordon, G. J. (2019). Environmental personhood. *Columbia Journal of Environmental Law, 43*(1), 49–91.

Harris, R. (2006). The transplantation of the legal discourse on corporate personality theories: from German codification to British political pluralism and American big business. *Washington and Lee Law Review*, 3(4), 1421–1478.

Miglani v State of Uttarakhand (2017) *Miglani v State of Uttarakhand (2017) Writ Petition (PIL) No. 140 of 2015 Uttarakhand High Court, Nainital.* https://indiankanoon.org/doc/92201770/

Montrul, S., Bhatt, R., & Girju, R. (2015). Differential object marking in Spanish, Hindi, and Romanian as heritage languages. *Language, 91*(3), 564–610.

Naff, J. N. (1972). Reflections on the dissent of Douglas, J., in Sierra Club vs. Morton. *American Bar Association Journal, 58*, 820.

Paredes Tamayo, I. (2024, August 16). A national park and its rangers in Bolivia endure persisting road construction, illegal mining. *Mongabay.* https://news.mongabay.com/2024/08/a-national-park-and-its-rangers-in-bolivia-endure-persisting-road-construction-illegal-mining/

Plurinational State of Bolivia (2008). *Law of the Rights of Mother Earth* (English translation). https://archive.wphna.org/wp-content/uploads/2014/07/2010-12-07-Bolivian-Law-of-rights-of-Mother-Earth.pdf

Pollman, E. (2012). Reconceiving corporate personhood. *Utah Law Review. 1629*, Available at SSRN: https://ssrn.com/abstract=1732910 or http://dx.doi.org/10.2139/ssrn.1732910

Reichel-Dolmatoff, G. (1997). Cosmology as ecological analysis: A view from the rainforest. In G. Reichel-Dolmatoff (Ed.), *Rainforest Shamans: Essays on the Tukano Indians of the Northwest Amazon* (pp. 7–20). Themis.

Santazilia, E. (2022). *Animacy and inflectional morphology across languages.* Brill.

Sleeper, A. (2023, May 11). How a tribal rights lawyer is winning back the rights of nature. *Resilience*. https://www.resilience.org/stories/2023-05-11/how-a-tribal-rights-lawyer-is-winning-back-the-rights-of-nature/

Stibbe, A. (2015). *Ecolinguistics: Language, ecology and the stories we live by*. Routledge.

Stone, C.D. (1972). Should trees have standing?–Towards legal rights for natural objects. *Southern California Law Review*, *45*, 450–501.

TNZPCC – The New Zealand Parliamentary Counsel Office. (2014). *Te Urewere Act 2014 No 51, Public Act: New Zealand Legislation*. Wellington.

Trumper, J. B. (2004). Current and historical problems in classification: Levels and associated themes, from the linguistic point of view. In G. Sanga & G. Ortalli (Eds.), *Nature knowledge: Ethnoscience, cognition, and utility* (pp. 68–94). Berghahn Books.

Turkewitz, J. (2017, September 26). Corporations have rights. Why shouldn't rivers? *The New York Times*. https://www.nytimes.com/2017/09/26/us/does-the-colorado-river-have-rights-a-lawsuit-seeks-to-declare-it-a-person.html

4

Preservation of the Environment and Development of Digital Technology

Hind Majdoubi

At a time when humanity is concerned about the sustainability of its development due to multiple challenges to overcome, the issue of environmental preservation remains relevant. Climate change has become increasingly threatening with its extreme phenomena, and the loss of biodiversity continues to grow as ecological balances risk being disrupted (Brondizio et al., 2019; IPCC, 2022). Despite the willingness of states and their cooperation on certain issues such as action on climate change and the agreement to achieve sustainable development goals, effective and timely solutions are slow to emerge. In this context, attempts and actions are multiplying to find solutions to correct or mitigate the problem of environmental damage. Among the solutions considered, digital technology is increasingly presented as an interesting alternative for the continuity of human relationships. Digital technology opens a virtual space that allows human activities to be carried out and is becoming an important reality with which we must now deal. It carries with it an evolution, even a revolution, which is changing the world. Digital technology can be apprehended in the context of what is called digital transformation. This is a transformation that has disrupted habits and ways of doing things that have been firmly established since time immemorial.

The digital transformation began in the 1990s with the advent of the internet for everyone (CESE, 2021). The trend quickly moved towards the creation of emails and websites for professionals, as well as the development of information searches through electronic search engines, the expression of opinions through forums or blogs, not to mention the explosion of social network registrations, etc. The full awareness of the potential of digital was rapid, but it must be acknowledged that the Covid-19 pandemic was the accelerator that marked the transition to a new era in the history of human relationships. Indeed, as part of measures taken to control the spread of the virus, populations were confined, which led to the cessation of human interactions in public spaces. Faced with a world at a standstill, the use of digital technology was able to provide, in almost all fields, a quick, adapted, and effective response to a major challenge. Taking the case of Morocco, for example, one can say that although the will to develop digital technology existed well before the pandemic, it was the pandemic that truly propelled its implementation.

Indeed, the responsibility of the state to ensure the continuity of public services was the reason behind the digitization of access and the use of these services in the context of e-administration. This acceleration of the digitization process can be explained by the fact that no visibility on the end of the pandemic was possible at the beginning of the health crisis. This is how development was carried out through the implementation of digital platforms, necessary for online health services, distance learning, e-justice, etc. Furthermore, companies used digital means to continue their activities and avoid losses due to confinement. Similarly, the confined population saw some of its segments switch to teleworking (work from home), take advantage of e-commerce, make online payments, and continue to lead a semblance of normal life by maintaining social relations at a distance. In this context, it has been noted that the use of digital technology has seen a clear increase during the pandemic period to ensure the continuity of virtual services. According to a document from the National Telecommunications Regulatory Agency of Morocco, the Moroccan Internet market increased between the end of March 2020 and the end of March 2021, with a penetration rate of around 85.11 percent, corresponding to 30.60 million subscribers.

In this context, highlighting the potential of digital technology to promote the development of human relationships, it becomes interesting to verify to what extent it has a positive or negative impact on the natural environment. To do so, it would be logical to question, on the one hand, the role and relevance of digital technology in relation to the environment, and on the other hand, the impact it has on the environment.

Close Links Between the Digital and the Natural Environment

The digital[1] transformation opens up a new horizon in which modes of human interaction do not require physical presence. In this context, it is useful to ask whether there are any links between the physical world and the virtual world of digital technology. To do this, we must first understand how the digital space can be apprehended in relation to the physical environment. It will then be interesting to understand why digital technology is often presented as an opportune solution for preserving the environment.

Digital technology has the potential to help protect the planet by promoting environmental sustainability, climate action, and pollution prevention (UN, 2022). The Global Environment Monitoring System for Air (GEMS Air), mentions (UN, 2022), is one of the first digital tools used by the United Nations Environment Programme (UNEP) to track the state of the environment in real-time at the global, national, and local levels. The digital initiatives leverage technology to halt the decline of the planet and accelerate sustainable finance, products, services, and lifestyles (DAVOS AGENDA, 2022). Experts say that in the years to come, a digital ecosystem of data platforms will be crucial to helping the world understand and combat a host of environmental hazards, from air pollution to methane emissions (Anderson & Caimi, 2022).

The digital transformation has opened up a new horizon in which modes of human interaction do not require physical presence. However, it is useful to ask whether there are any links between the physical world and the virtual world of digital technology. To do this, we must first understand how the digital space can be apprehended in relation to the physical environment. It will then be interesting to understand why digital technology is often presented as an opportune solution for preserving the environment (Coalition for Digital Environmental Sustainability (CODES), 2022).

Understanding Digital Space's Role in Environmental Preservation

Digital technology can be understood as a medium that allows access and interaction in a dematerialized environment where the existence of humans and their relationships with each other are purely virtual. This situation raises questions about the classification of the immaterial environment of digital technology with reference to the classic physical environment. To address this, it is possible to consider this immaterial environment either as a parallel dimension or even opposed to the physical environment, or as a new extension of the environment through an expansion of its definition. Before that, it is useful to identify any similarities between the physical environment and the immaterial digital environment. In this sense, we can observe that the similarity between the natural environment and the digital environment is revealed at least at two levels. Firstly, they are both worlds that escape the artificial division created by humans, even if human intervention exists to establish state sovereignty. Generally, environmental protection ignores the artificial boundaries set by humans and calls for a global and effective cooperation of all states to achieve the result of protection. Similarly, the Internet has transformed the world into a global village. Despite the power of states to control or limit access to the Internet, it remains a space for virtual human interaction, with its limits primarily determined by access to the network. Secondly, the similarity between the physical and digital environments is evident in their function as a support for human activities, including human relationships within a context that facilitates the development of those activities.

Given the above, a return to the definition of the environment is inevitable, despite the difficulty of precisely defining the term. In this regard, in the first step, we will adopt the definition of the term "*environner*" by Paquot (2010), which generally refers to the notion of surrounding, encircling, and enclosing. He then specifically qualifies the environment as "the relationships between humans and the environment in which they live." In another definition, the environment refers to "the set of natural or artificial (physical, chemical, and biological) and cultural (sociological) conditions in which living organisms develop (including humans, animal, and plant species)."[2] From these definitions, two interesting elements can be identified in relation to the environment. Firstly, the element refers to the set of natural or artificial and cultural conditions, and secondly, it relates to the development of living organisms. In other words, is it possible to consider the immaterial space of the digital world as an artificial condition that constitutes a living environment for humans? Indeed, the digital world offers the possibility for humans to have an existence and to establish and maintain virtual relationships that are recognized in the physical world. At this stage, it is not necessary to engage in philosophical debates about the distinction between life and existence (Jacquet, 2011). We will then only focus on the general meaning of existence, which is defined by the Cambridge dictionary[3] as "the state of being real, or being known or recognized." The appreciation of the digital world in relation to the environment always maintains the dilemma of whether to consider the digital as a world or even a parallel or opposite dimension of human life, offering humans the exclusivity of a virtual existence, or rather to consider it as a component of the extended environment in a general sense. If we opt for the first path, the digital environment will be opposed to the ecological, physical environment in the sense that the former could not be a support for life in terms of a biotope. However, if we choose the second option, which refers to "all the natural or artificial and cultural conditions in which living organisms develop," then the definition of the environment could be extended to encompass, through the artificial conditions of the digital, a virtual environment allowing for the development of humans. In this sense, the environment understood as a medium would include the following physical environments: solid environments (the lithosphere), liquid environments (the hydrosphere), and aerial environments (the atmosphere), and would extend to a new environment, namely a digital or virtual existence environment. As such, we can observe an expansion of the notion of environment to include the physical medium understood as a medium of life and an immaterial medium understood as a medium of existence.

In any case, the success of the digital is mainly justified by the potential it offers for human development. It is curious, then, to understand what opportunities it has for reducing pressure on the natural physical world.

Digitalization: Opportunities for Humans, Less Pressure on Environment

The dematerialization of human activities through the use of digital technology has expanded opportunities for development in all aspects of human life. It is important to recognize that the advantages of digital technology lie in the possibility it offers to expand the boundaries of human existence. In this sense, humans can lead a material existence as well as a virtual one. Two worlds, namely the physical and virtual worlds, coexist, compete, and even complement each other. If the virtual world is successful, it is precisely because it allows us to escape the boundaries of the physical world and creates opportunities for relationships that would not have been possible to imagine due to the limits of the material world.

Thanks to its characteristics of global reach and immediacy, there is no longer a need to travel to go to work or attend meetings. Mail is dematerialized, markets are virtual, and it offers opportunities for planetary relationships. Public services are increasingly accessible at all times, not to mention being transparent and fast, etc. The advantages of digitalization can also be counted in the fact that they allow for the reduction of costs incurred by physical travel, cutting distances, and optimizing time.

On another level, digital technology has also developed thanks to the possibilities of access and storage of information (e.g., information search has become easier, information content can be shared and modified, in real-time, by several people without the need for physical presence together, the case of data centers is also interesting in terms of storage and mobilization of information). In addition, the possibilities of very precise and rapid processing of information through the use of artificial intelligence have revolutionized the approach to human work, since, for example, certain tasks can be digitized and automated to optimize costs and time. In short, digital transformation is explained by the changing socio-economic-cultural models of humans, both in their private and professional relationships, as well as in their relationships with public administration.

From all the above, it is worth asking whether the digital opportunities mentioned earlier also bring benefits to the environment. In this regard, it can be quickly stated that savings can be made on costs related to travel, logistics, etc., which would ostensibly avoid the pollution generated by these physical activities. This idea was reinforced during the general confinement period, when an improvement in environmental indicators was observed. In this sense, a reduction in nitrogen dioxide concentrations of up to 60 percent was observed in some European cities (Rouïl & Colette, 2021). Additionally, digital technology enables better environmental protection by exploiting its potential for observation, data collection, modeling, etc. This can improve the energy and hydraulic efficiency of systems in certain sectors, both in terms of production and use.

This naturally leads us to consider digitalization as a solution to reduce pressures on the environment and even achieve sustainable development. The potential of digital technology has even been presented as an opportunity for environmental preservation and sustainable development. The latter relies on the indivisibility of economic, social, and environmental dimensions for the long-term development of humanity. In this sense, we will mention three examples of this position that extols the benefits of digital transformation.

The first example comes from the Resolution adopted by the United Nations General Assembly, establishing the 17 Sustainable Development Goals for global sustainable development by 2030. In fact, we can read the following in the Resolution: "The expansion of computing and communications and the global interdependence of activities have the potential to accelerate human progress, reduce the digital divide, and create knowledge societies, not to mention scientific and technological innovation in fields as diverse as medicine and energy" (General Assembly, 2015). Furthermore, some of the Sustainable Development Goals promote this transformation. For example, Sustainable Development Goal 9 aims to build resilient infrastructure, promote sustainable industrialization that benefits everyone, and encourage innovation. It calls for "substantially increasing access to information and communications technology and striving to provide universal and affordable access to the Internet in least developed countries by 2020" (General Assembly, 2015).

The second illustration is taken from a document published by UNEP in 2022. The document outlines "the vision for UNEP's digital transformation over the next four years, in response to the changing digital development landscape and the evolving needs of their partners, including government, businesses, finance, and civil society" (UNEP, 2022). Additionally, the document states that UNEP's digital transformation approach should not merely involve digitalizing its products and services. Instead, it should bring transformational changes in five areas: environmental decision-making, economic incentives and business models, human behaviors, and environmental governance.

The third illustration is from a communication issued by the European Commission, which presents a European Green Deal for the European Union (EU) and its citizens. The European Green Deal aims to transform the EU into a fair and prosperous society, with a modern, resource-efficient, and competitive economy where there are no net emissions of greenhouse gases in 2050 and where economic growth is decoupled from resource use (European Commission [EC], 2012). The communication highlights that digital technologies are a critical enabler for achieving the sustainability goals of the Green Deal across various sectors. Therefore, the Commission intends to explore measures to ensure that digital technologies can accelerate and maximize the impact of policies to address climate change and protect the environment. The communication also notes that digitalization presents new opportunities for distance monitoring of air and water pollution, as well as monitoring and optimizing how energy and natural resources are used (EC, 2012).

Of the three selected illustrations, digital technology is presented as a sustainable solution, and therefore, it serves sustainable development and, by extension, is reconciled with the objective of environmental preservation. However, considering digital technology solely from the perspective of its potential as a facilitator or even an accelerator of sustainable development, relying on the aspect of environmental preservation, is actually reducing its overall impact on the environment. Therefore, it is necessary to thoroughly verify whether the use of digital technology is genuinely reconciled with the objective of environmental preservation.

Sustainability of Digital: Awareness and Improvement Needed

Digital Pollution—Silent and Threatening Pollution

If we examine the question of the sustainability of digital technology, there is often a tendency to think that the use of digital services takes place in a virtual world and therefore has no connection to the physical world. Furthermore, its sustainability is explained by the fact that bypassing the physical world allows for the elimination of costs associated with time and distance, and as a result, also eliminates the impacts of physical activities on the environment.

Indeed, digital technology, a relatively recent development, is beginning to reveal its flaws, among which are its negative effects on the environment. In this regard, it has been proven that it causes real pollution and impacts the environment and threatens to do so even more seriously in the future, as the trend of digital pollution is accelerating (Shift Project, 2021). Its pollution is increasingly denounced (Flipo, 2017; Pitron, 2020). Its seriousness lies in the fact that, although it appears abstract and silent, its impact is very real and physical. In this sense, Flachaire (as cited in Patard, 2023) has described the situation well: "We have the impression that a click remains abstract, that everything happens somewhere in the cloud. But the use of new technologies and servers has a very heavy weight on the environment." In reality, digital technology is not actually immaterial, and its environmental sustainability is actually a myth, which raises questions about its impact on the physical environment.

While it allows access to a virtual environment for human activities, the supports used to access it are still physical (computers, smartphones, screens, chargers, servers, not to mention millions of kilometers of copper cables and fiber optics, etc.). In addition, the consumption of energy necessary to operate it and the use of water resources cause considerable pressure on the environment. In this sense, a study has asserted that the digital world's environmental footprint and its evolution between 2010 and 2025 (GREEN IT, n.d.) are not neutral in terms of humanity's overall footprint. The study found that the digital world's primary energy consumption is 4.2 percent, its contribution to greenhouse gas emissions is 3.8 percent, and it consumes around 0.2 percent of global water resources. Additionally, the use of digital technology for storing and exchanging data has an environmental impact in terms of energy consumption and greenhouse gas emissions (Berthoud, 2017).

Therefore, the negative impact of digital technology on the environment can be observed at least at three levels. Firstly, it is evident in the production of digital equipment and infrastructure necessary for its use. In this context, the pressure on the environment is caused by the exploitation of natural resources, sometimes non-renewable, and ironically, greater than the weight of the finished product. The French Agency for Ecological Transition (ADEME) specifies that "it takes 50 to 350 times their weight in materials to produce high-tech electronic devices" (ADEME, 2021). Secondly, digital technology has a negative and dual impact on the environment during its operation. On the one hand, energy and water resources are required for its operation, and on the other hand, greenhouse gas emissions generated by its use. In this sense, ADEME further specifies that "the manufacture of a 2 kg computer represents 800 kg of mobilized raw materials and 124 kg of CO2 generated, out of the 169 kg emitted throughout its life cycle" (ADEME, 2021). In third place, the negative impact of digital technology on the environment is also revealed by the waste generated at the end of life of digital products, which causes significant pollution.

It should be noted that while the figures may seem low compared to the pressure from physical human activities, they should not be underestimated. In fact, the growth of the digital phenomenon is expected to increase its ecological footprint in the future, due to changes in lifestyles. In this context, the International Energy Agency is warning humanity about the growing concerns related to the future global environmental impacts of digital technology. Indeed, through the rebound effect, environmental gains are lost through the multiplication of digital uses and users. (EcoInfo, n.d.)

Re-examining Digital Sustainability

The particularity of digital technology that is very appealing lies in the fact that it allows humans to communicate and even establish legal relationships without the need to be physically close. It has been nicely presented by Rénier et al.

(2022) as offering "proximity at a distance" to describe the experience of French farmers who have used it to communicate about their agricultural practices and overcome physical distance, which can be a source of different weather and geographical conditions.

It is clear, let us recall, that during the Covid-19 pandemic, the use of digital technology has provided the opportunity to maintain human relationships in a context of confinement. Indeed, the switch to online education, the development of e-government to continue public services, teleworking which saved jobs and businesses, and the success of social networks are all illustrations that have allowed digital users to maintain a semblance of social life, despite physical distance.

On another level, it should also be noted that the use of digital technology presents an interesting solution for avoiding costs generated by, for example, travel time, transportation, and logistics, while also reducing the impact of these activities on the environment.

From this perspective, digital technology appears to contain enormous potential for the development of human activities and is therefore very appealing to both businesses and public decision-makers, among others. In this regard, we can read in the report of the special commission for the new model of development that it considers digital technology to be "a real lever for change and development." Furthermore, it recommends "giving it special attention at the highest level of the state as a catalyst for structural and high-impact transformations" (Commission Spéciale sur le Modèle de Développement [CSMD], 2021, p. 151).

This highlights the enormous development potential of digital technology, which ultimately explains the phenomenon of digital transformation. A phenomenon that is now well established, poised to accelerate, and therefore bound to last, without the possibility of a return to the past (Organisation de Coopération et de Développement Économiques [OCDE], 2021, p. 1) [4].

In this context, it is useful at first to demystify the intangibility of the digital in people's minds in order to become aware of its material impacts on the environment (Sibo, 2020). Also, it is important to understand that ensuring the true sustainability of digital is crucial for correcting its negative impacts on the environment. To achieve this, the use of law becomes essential for the regulation of digital. Indeed, it is necessary to define the situations of negative impact of digital and hold those responsible for them (Djazouli-Bensmain, 2019). The sustainability of digital should not be limited to its ability to circumvent the physical world and therefore the physical activities that put pressure on the environment. In this context, extended responsibility seems capable of legally holding manufacturers of digital supports accountable, by extending their responsibility from the design stage to the end-of-life treatment of these supports. It should be noted that some legal systems have already recognized extended responsibility, such as the European Union, Japan, China, etc. However, other states have only provided for it as a principle without necessarily applying it.

This is the case in Morocco, for example[5]. Extended responsibility is interesting for moving digital towards a circular economy through legal principles such as the polluter pays principle (Eude, 2019). The approach consists of removing waste from its status as something that no one wants and that causes pollution, towards a paradigm that recognizes its value and encourages its reuse by integrating it into the industrial manufacturing loop.

On another level, since digital impacts the environment due to its energy and water needs, its functioning must be reconsidered by holding its manufacturers and users accountable (Pisani-Ferry, 2022). To do this, waste reduction must be prioritized and recycling and renewable solutions must be favored, even if they are more expensive. Similarly, the environmental impact of digital must be anticipated, and for this, there must be systematic evaluation of the environmental relevance of connected technologies, according to Shift Project (2020).

Finally, in order to preserve the environment, states have committed to reducing greenhouse gas emissions. It then becomes useful and reasonable to take into account the air pollution caused by digital. In this sense, the law must aim for carbon neutrality of digital and impose transparent communication to the public in order to encourage technological innovation in green digital.

Conclusion

As a conclusion, the preservation of the environment and the development of digital technology are two important but sometimes conflicting goals. Additionally, digital transformation reflects a new reality in the way human relationships are conducted, ensuring closeness in distance and offering instantaneity in time. Digital transformation has not yet delivered its full potential for development and growth. Continuous innovations in new technologies, combined with the challenge of digital inclusion, are factors that already point to an exponential trend in the development of digital media. This development is closely related to the natural physical environment at all stages of the life cycles of digital media, leading ultimately to a shared destiny. It is therefore urgent to take into account the environmental impact of digital media in order to ensure its sustainability and prevent digital transformation from becoming ephemeral.

Notes

1 According to the Collins Dictionary, "*digital*" means "*of, relating to, resembling, or possessing a digit or digits,*" "*performed with the fingers,*" "*representing data as a series of numerical values,*" "*displaying information as numbers rather than by a pointer moving over a dial,*" "(electronics) responding to discrete values of input voltage and producing discrete output voltage levels, as in a logic circuit," etc. The source of this information can be found at: https://www.collinsdictionary.com/dictionary/english/digital
2 https://www.dictionnaire-environnement.com/environnement_ID1045.html
3 https://dictionary.cambridge.org/fr/dictionnaire/anglais/existence
4 From then on, countries are faced with a major challenge. It is unlikely that economies and societies will return to the models of the "pre-COVID-19" era: the crisis has demonstrated the striking potential of digital technologies, and some developments may be too profound for a return to the past to be considered.
5 Article 2 of the Framework Law 99-12 forming the national charter for the environment and sustainable development calls for an update of the legislative framework governing waste, by integrating the principle of extended responsibility in particular, Official Bulletin n 6240 of the March 20, 2014.

References

ADEME (2021). *La face cachée du numérique, Réduire les impacts négatif du numérique sur l'environnement* [The hidden side of digital technology: Reducing its negative impacts on the environment]. https://cnm.fr/wp-content/uploads/2021/08/ademe_guide-pratique-face-cachee-numerique.pdf. Retrieved from https://presse.ademe.fr/wp-content/uploads/2022/01/DP_Numerique-responsable-190122.pdf

Anderson, J., & Caimi, G. (2022, January 10). *A three-part game plan for delivering sustainability digitally: Digital technologies can do much more than measure progress to net zero. They can support your entire sustainability strategy.* Retrieved from https://www.bain.com/insights/a-three-part-game-plan-for-delivering-sustainability-digitally/

Berthoud, F. (2017). Numérique et écologie [Digital and Ecology]. *Annales des Mines - Responsabilité et Environnement, 87*, 72–75. https://doi.org/10.3917/re1.087.0072

Brondizio, E. S., Settele, J., Díaz S., & Ngo, H. T. (Eds.). (2019). *Global assessment report on biodiversity and ecosystem services of the Intergovernmental Science-Policy Platform on Biodiversity and Ecosystem Services.* IPBES Secretariat, p. 1148. https://doi.org/10.5281/zenodo.3831673

Coalition for Digital Environmental Sustainability (CODES). (2022). *Action Plan for a sustainable planet in the digital age.* Retrieved from https://wedocs.unep.org/bitstream/handle/20.500.11822/38482/CODES_ActionPlan.pdf?sequence=3&isAllowed=y

Commission Spéciale sur le Modèle de Développement (CSMD). (2021). *Commission Spéciale sur le Modèle de Développement, Le nouveau modèle de développement, Rapport général* [Special Commission on the Development Model, The New Development Model, General Report towards a responsible and inclusive digital transformation]. Retrieved from: https://www.csmd.ma/documents/Rapport_General.pdf

Conseil Economique, Social et Environnemental (CESE). (2021). *Avis du Conseil Economique, Social et Environnemental Vers une transformation digitale responsable et inclusive* [Opinion of the Economic, Social and Environmental Council Towards a responsible and inclusive digital transformation]. Retrieved from https://www.cese.ma/media/2022/01/Avis-Transformation-digitale-VF-.pdf

Conseil Economique, Social et Environnemental (CESE). (2021). *Vers une transformation digitale responsable et inclusive* [Towards a responsible and inclusive digital transformation]. Retrieved from https://www.cese.ma/media/2023/01/Transformation-digitale-VF.pdf

DAVOS AGENDA. (2022, January 19). *3 ways digital technology can be a sustainability game-changer.* Retrieved from https://www.weforum.org/agenda/2022/01/digital-technology-sustainability-strategy/

Djazouli-Bensmain, A. (2019). *Le déchet numérique: objet juridique non identifié* [E-waste: Unidentified legal object]. *Droit et Ville, 87*, 147–160. https://doi.org/10.3917/dv.087.0147

EcoInfo. (n.d). *Effets rebond du numérique* [Digital rebound effects]. Retrieved from https://ecoinfo.cnrs.fr/thematiques/consequences-du-numerique/effets-rebond-du-numerique/

Eude, M. (2019). L'économie circulaire, de la notion économique aux principes juridiques complexes [Circular economy, from economic concept to complex legal principles]. *Droit et Ville, 87*, 291–307. https://doi.org/10.3917/dv.087.0291

European Commission [EC]. (2012). *Communication from the Commission to the European Parliament, the Council, the European Economic and Social Committee and the Committee of the Regions, A blueprint to safeguardd Europe's water resources.* Brussels, 14.11.2012, COM (2012) 673 final. Retrieved from http://eur-lex.europa.eu/LexUriServ/LexUriServ.do?uri=CELEX:52012DC0673:EN:NOT

Flipo, F. (2017). Peut-on croire aux TIC « vertes » ? [Can we believe in "green" ICT?]. *Annales des Mines - Responsabilité et Environnement, 87*, 105–107. https://doi.org/10.3917/re1.087.0105

General Assembly. (2015). *Resolution A/70/L.1, Transformer notre monde: le Programme de développement durable à l'horizon 2030, (25 septembre 2015)* [Resolution A/70/L.1, Transforming our world: The 2030 Agenda for Sustainable Development]. Retrieved from https://unctad.org/system/files/official-document/ares70d1_fr.pdf

GREEN IT. (n.d). *Empreinte environnementale du numérique mondial* [Global digital environmental footprint]. Retrieved from https://www.greenit.fr/etude-empreinte-environnementale-du-numerique-mondial/

Jacquet, F. (2011). Vie et existence: vers une cosmologie phénoménologique [Life and existence: Towards a phenomenological cosmology]. *Les Études philosophiques, 98*, 395–419. https://doi.org/10.3917/leph.113.0395

Intergovernmental Panel on Climate Change (IPCC). (2022). *Climate Change 2022: Mitigation of Climate Change—Summary for Policymakers.* Retrieved from https://reliefweb.int/report/world/climate-change-2022-mitigation-climate-change-summary-policymakers?gclid=Cj0KCQjwk7ugBhDIARIsAGuvgPZnw71Yfjgq6C7aEt7BOSLXMtGfqALvsmbsxXBUXkD4DsbN72ZzCdEaAhwhEALw_wcB

Organisation de Coopération et de Développement Economiques (OCDE). (2021). *La transformation numérique à l'heure du COVID-19 : Renforcer la résilience et combler les fractures, Supplément à l'édition 2020 des Perspectives de l'économie numérique* [Digital transformation in the time of COVID-19: Strengthening resilience and bridging gaps, Supplement to the 2020 edition of Digital Economy Outlook] OCDE, Paris. Retrieved from https://www.oecd.org/fr/numerique/transformation-numerique-covid.pdf

Paquot, T. (2010). *Environnement » et « milieu(x) urbain(s) », enquête étymologique* ["Environment" and "urban environment(s)," etymological investigation]. In T. Paquot (Ed.), *Philosophie de l'environnement et milieux urbains* (pp. 19–42). La Découverte.

Patard, A. (2023). *Impact du digital sur l'environnement: enjeux et solutions pour un web plus propre* [Impact of digital on the environment: Challenges and solutions for a cleaner web.]. Retrieved from https://www.blogdumoderateur.com/impact-digital-environnement-enjeux-solutions-web-plus-propre/

Pisani-Ferry, J. (2022, December 2). *Action climatique, enfin les ennuis commencent, Terra nova* [Climate action, trouble begins, Terra Nova.]. Retrieved from https://tnova.fr/ecologie/transition-energetique/action-climatique-enfin-les-ennuis-commencent/

Pitron, G. (2020). La guerre des métaux rares ou la face cachée de la transition énergétique et numérique [The rare metals war or the hidden side of the energy and digital transition]. *Annales des Mines – Responsabilité et Environnement, 99*, 74–75. Retrieved from https://www.cairn.info/revue-responsabilite-et-environnement-2020-3-page-74.htm?contenu=auteurs

Rénier, L., Cardona, A., Goulet, F., & Ollivier, G. (2022). La proximité à distance: Comment les agri-youtubeurs communiquent sur leurs pratiques [Distance proximity: How agri-youtubers communicate about their practices]. *Réseaux, 231*, 225–257. https://doi.org/10.3917/res.231.0225

Rouïl, L., & Colette, A. (2021). Quels effets du confinement sur la qualité de l'air: retour sur le printemps 2020. [What are the effects of confinement on air quality: A look back at spring 2020]. *Environnement, Risques & Santé, 20*, 308–316. https://www.cairn.info/revue--2021-3-page-308.htm.

Shift Project. (2020). *Déployer la sobriété numérique, résumé aux décideurs* [Deploying digital sobriety, summary for decision-makers]. Retrieved from https://theshiftproject.org/wp-content/uploads/2020/10/Deployer-la-sobriete-numerique_Resume_ShiftProject.pdf

Shift Project. (2021, March 26). *Impact environnemental du numérique: Tendances à 5 ans et gouvernance de la 5g, Synthèse* [Environmental impact of digital: 5-year trends and governance of 5g, Synthesis.]. Retrieved from https://theshiftproject.org/article/impact-environnemental-du-numerique-5g-nouvelle-etude-du-shift/

Sibo, C. (2020). Immatérielle, l'expansion mondiale des TIC ? ["Immaterial," the global expansion of ICT?]. In C. Leterme (Ed.), *Impasses numériques: Points de vue du Sud* (pp. 105–117). Éditions Syllepse.

UN. (2022, August 17). *How digital technology and innovation can help protect the planet.*

UNEP. (2022). *Digital transformation* [Digital transformation]. Retrieved from https://www.unep.org/resources/policy-and-strategy/digital-transformation-becoming-innovative-agile-and-collaborative

5

The Law of Sustainable Development and Hazardous Waste

Khadija Anouar

A healthy environment is indispensable for the enjoyment of human rights. As a source of pollution, the management and disposal of waste have gained increasing significance due to economic development, urban expansion, and contemporary lifestyles that amplify the demand for packaging and disposable products. In France, the issue of waste is not a recent phenomenon; the first measure was implemented by Eugène Poubelle in 1883 (Cirelli & Maccaglia, 2019). According to Algerian law, hazardous waste is regulated by law n° 19/01 of December 12, 2001, which pertains to the management, control, and elimination of waste.

In Morocco, the initial legislation governing waste was the Dahir of 1914, which addressed unhealthy, inconvenient, or dangerous establishments. However, specific legislation on the subject was only adapted in 2006 with Law No. 28-00, which pertains to the management of waste and its elimination in general terms.

Waste, as defined by the provisions of Article 3 of the law on waste, is any residue resulting from a process of extraction, exploitation, transformation, control, use, or filtration in general, and any material that must be properly eliminated by its producer to avoid its harmful effects.

This topic is of paramount interest because Morocco and Algeria are currently experiencing a steady increase in the volume of hazardous waste. This increase is attributable to the growth in economic activity, population, and particularly urban expansion, as well as changes in the lifestyles of citizens. Waste poses a significant threat not only to the environment but also to human health, making the management and disposal of hazardous waste a strategic priority for the State.

Thus, in 2006, Law No. 28-00 was promulgated relating to the management of waste and its elimination[1]. Said law divided waste into several categories including dangerous ones.

The topic of hazardous waste management is indeed of paramount importance, prompting us to pose the following question: How does Law No. 28-00 facilitate the management of hazardous waste, and how effective is it?

To address this question, our analysis will initially explore the benefits offered by the law through the management procedure it assigns to the said waste. Subsequently, we will scrutinize the shortcomings of this legislation and propose potential enhancements in this domain.

Legal Mechanisms of Hazardous Waste Management

The components of hazardous wastes are indeed highly diverse in terms of their chemical composition and physical design. These characteristics render them susceptible to undeniable effects on human health and the environment in which they are produced, disposed of, or handled.

The Legal Status of Hazardous Waste Management

Morocco has strengthened its environmental legislation through the adoption of Law No. 28-00 on waste management and disposal, especially with regard to hazardous waste that negatively impacts the environment and public health. Other types of waste, such as household, medical, or industrial waste, may contain objects or substances belonging to the category of hazardous waste such as batteries, used oils, and sharp or infected objects, while hazardous waste may include all other types of waste. Law No. 28-00 established the modalities and norms for the management of all types of waste including hazardous waste. Firstly, it prohibits the mixing of hazardous waste with other non-hazardous types of waste upon collection in order to avoid infections. It also establishes the obligations of producers and holders of this type of waste for their disposal. Hazardous waste according to the legislation is any form of waste that, due to its hazardous, toxic, reactive, explosive, flammable, biological, or bacterial nature, constitutes a danger to the ecological balance as set out in the international standards in this field.

The legal provisions of Article 58 of Law No. 28-00 state in essence that the operation of treatment, recovery, incineration, storage, disposal, or landfill facilities for hazardous, industrial, or medical and pharmaceutical waste requires the deposit of a financial guarantee. This financial guarantee is intended, if necessary, to cover possible interventions in the event of accidents occurring before or after closure, as well as to maintain the safety of the installation and the surveillance of the site (World Bank, n.d.).

However, the said financial guarantee is in no way intended to cover compensation owed by the operator to third parties who may suffer harm as a result of pollution or accidents caused by the facility (Dahchour & Hajjaji, 2020). The management of hazardous waste is carried out in specialized treatment facilities subject to authorization regimes and which are designated by the administration in accordance with the provisions of the national master plan for hazardous waste, which is a plan that is established by the administration in coordination with the local authorities over a five-year period (Hansen, 2023). Management of hazardous waste facilities is important. Thus, we need to look into the facilities where these wastes are managed as well as the sanctions in case of failure to comply with the regulations in force.

According to Law No. 28-00, two types of facilities are in question: the first is of a general nature, the controlled landfills, and the second is of a special nature, the installations for the treatment, storage, recovery, and elimination of waste, especially of a hazardous nature.

A controlled landfill is defined as a facility or site whose characteristics are identical to a permanent waste deposit. It classifies its waste according to the nature and characteristics of the waste. Category n°1 is that of household and similar waste, category n°2 is that of non-hazardous industrial, medical and pharmaceutical waste, and finally, agricultural, and inert waste. Category n°3 is hazardous waste.

The opening, transfer, substantial modification, and closure of landfills are subject to authorization from the administration after a public inquiry and the agreement of the Wali of the region as well as that of the governor of the province or prefecture of the landfill location.

Waste management requires treatment, recovery, storage, and disposal of waste, which implies having special facilities for these purposes. The opening, transfer, modification, and closure of a facility are also subject to authorization. The disposal of hazardous waste must respect the modalities set by Law No. 28-00. Also, there are specific places designated for this purpose, otherwise, penalties shall be imposed.

Hazardous Waste Management Process

According to the legal provisions of Article 29 of Law No. 28-00 on waste management and disposal, waste management is defined as any operation of pre-collection, collection, storage, transportation, landfilling, sorting, treatment, recovery, recycling, and disposal of waste, including the control of these operations as well as the monitoring of the facility during operation after its closure. The management of hazardous waste is carried out in facilities subject to the authorization scheme issued by the administration and the authorities concerned, in accordance with the national master plan for the management of hazardous waste. The legislation in this area is based on the said law and the decree n°2-14-85 of January 20, 2015, on the management of hazardous waste.

The transport of waste is subject to strict regulations, where appropriate for hazardous waste in view of its harmfulness. Therefore, hazardous waste must be transported in special containers according to its specificities and classifications and can be transported by the waste producer or holder himself or by a collector-transporter authorized for this purpose to special facilities to eliminate or recover it. A tracking slip must accompany the transport of these wastes; unless the sender has the agreement of the receiver for the reception of these wastes, no transport can be done . As for the administration, it is obliged to record in a register the information recorded for a period of five years from the date of recording, including the type, quantity, nature, and origin of the hazardous waste. In this sense, Article 30 of Law No. 28-00 stipulates that the collection and transport of hazardous waste are subject to authorization by the administration. This authorization is granted for a maximum period of five years and can be renewed. It is granted only after the following conditions are met: the entity must undertake to carry out, on a primary basis, the activities of collection and transport of hazardous waste; possess sufficient financial capacity to execute these activities; employ qualified and trained personnel; commit to implementing preventive and sanitary measures to ensure the safety of the personnel; and be equipped with the appropriate equipment for the collection and transport of hazardous waste.

The transport of hazardous waste from the production site can only be carried out if the packaging and containers necessary for their transport are clearly and visibly labeled with the waste type, in accordance with the standards in force (Article 31 of Law No. 28-00). The transport of hazardous waste must be accompanied by a tracking slip containing information on the sender, the carrier, the recipient, the nature and quantity of the waste, the mode of transport, and the modalities of its disposal. In Algerian law, according to the ministerial decision of 2 September 2013 specifying the documents of the file for the authorization of special transport of hazardous waste, it is forbidden to bury, throw, store or deposit hazardous waste in places other than the facilities dedicated to them in accordance with the provisions of the legislation.

Thus, any natural or legal person who deposits, or causes to be deposited, hazardous waste with an unauthorized natural or legal person is jointly and severally liable to that person for any damage caused by that waste. During collection, transport, storage, recovery, disposal, or landfill operations, hazardous waste may not be mixed with other categories of waste. However, the administrator may grant a derogatory authorization to the facilities concerned when the mixing of hazardous waste with other waste is necessary for the recovery, treatment, or disposal of such waste (Article 35 of Law No. 28-00).

In addition, any natural or legal person who produces, collects, transports, stores, or disposes of hazardous waste must have an insurance contract covering their professional liability (Article 36 of Law No. 28-00). According to Article 37, the producers of hazardous wastes and the persons holding the authorizations shall keep a register in which they shall record the quantities, type, nature and origin of the hazardous wastes that they have produced, collected, stored, transported, recovered or disposed of, and shall communicate annually to the Administration the information of this type corresponding to the past year. This register is subject to inspection by the administration.

As for the control and observation of the committed infringements that are carried out by the agents and officers of the judicial police, as well as the civil service and agents commissioned for this purpose by the administration, they have the possibility to request from the operator of the dump or the facility all the necessary information. The administration, in its turn, can request private expertise to evaluate the degree of harmfulness of these wastes. Law No. 28-00 foresees administrative and penal sanctions. Administrative sanctions are sanctions that the administration can impose under the law. Article 6 of the said law states that any person who holds or produces waste must dispose of it or have it disposed of taking into account the preservation of human health and the environment.

In addition to the administrative sanctions, penal sanctions are also applied. These can take the form of fines exceeding 1200 MAD, imprisonment, or both. These penalties are enforced in cases where the holder of the hazardous waste disposes of it or buries it in inappropriate locations, fails to provide information, or provides incorrect information to the administration, or obstructs the duties of a control agent.

The Main Actors in Hazardous Waste Management or Planning

The primary stakeholders in waste management can vary depending on the type of waste, but generally, local authorities, communes, and the private sector are always involved, especially in the management of hazardous waste. The municipality establishes a strategy for the management of the public service using its own human and financial resources. The management is done by deliberation of the communal council or of the grouping of communes of a certain public service.

In the case of delegation, the commune entrusts the management of a public service to a private or public delegate. It is a remunerated delegation that aims to improve the quality of the services provided and access to funding. In the event of an infringement, the responsibility of the municipality and the delegate will be jointly retained if the case in question reveals an imputable fault; otherwise, only the delegated party will be held responsible. It should be noted that the delegation is subject to conditions.

According to Article 9 of the law, the national master plan for hazardous waste management is established by the administration in collaboration with the local authorities and professionals concerned. It must determine the objectives to be reached in terms of collection and elimination rates of hazardous wastes, as well as the appropriate sites for the implementation of elimination and storage facilities for these wastes, taking into account the places where they are produced and the urban planning documents, adding to this a five- and ten-year forecast inventory of the quantities of wastes to be collected or eliminated according to the nature, type and origin, then an investment program for the same duration as the inventory, including the costs of building storage and disposal facilities, waste recovery and recycling, not forgetting the financial and human resources required and the measures to be taken in terms of information, awareness and advice. This plan is established by decree for a period of ten years and can be revised when need be.

In addition, the territory of each region must be covered by a regional master plan for the management of non-hazardous industrial, medical, and pharmaceutical waste, as well as final agricultural and inert waste. This plan determines, in particular, the objectives to be achieved in terms of collection and disposal rates for these categories of waste. It also identifies appropriate sites for the establishment of disposal and storage facilities, taking into account the guidelines of urban planning documents. Furthermore, the plan includes a five-year and ten-year forecast inventory of the quantities of waste to be collected and disposed of according to their origin, nature, and type. An investment program for the same period is outlined, which includes the evaluation of the costs associated with creating controlled landfills and facilities for the treatment, storage, recycling, or recovery of this waste, as well as the rehabilitation of uncontrolled landfills. Additionally, the plan specifies the necessary financial and human resources and outlines measures to be taken in terms of information dissemination, awareness-raising, and advisory services.

The regional master plan is drawn up by the regional council and under the responsibility of the Wali (governor), in consultation with a consultative commission made up of representatives of the prefectural and provincial councils, the administration, professional bodies concerned with the production and elimination of these wastes, and environmental protection associations in the region concerned.

The master plan takes into account the needs and potentials of neighboring areas outside the area of implementation, as well as the possibilities of inter-regional cooperation in this field. The plan is subject to a public inquiry. Following the Algerian legislation, according to Article 2 of the implementation decree n° 03/477 of 9 December 2003, the Minister in charge of the environment is responsible for establishing the plan for the management of hazardous waste for a period of ten years. In Morocco, the territory of each prefecture or province must be covered by a prefectural or provincial master plan for the management of household and similar waste.

This plan determines, in particular, the objectives to be reached in terms of collection and elimination rates for household and similar waste. It identifies appropriate sites for the installation of disposal and storage facilities for this waste, taking into account the guidelines of urban planning documents. Additionally, the plan includes a five-year and ten-year forecast inventory of the quantities of waste to be collected and disposed of according to their origin, nature, and type. An investment program for the same period is outlined, which includes the evaluation of the costs associated with creating controlled landfills and facilities for the treatment, recovery, storage, or disposal of this waste, as well as the rehabilitation of uncontrolled landfills. Furthermore, the plan specifies the necessary financial and human resources and outlines measures to be taken in terms of information dissemination, awareness-raising, and advisory services.

Hazardous Waste Law and Sustainable Development

Hazardous wastes are wastes that can cause substantial harm to human health and safety or to the environment when improperly handled. Hazardous wastes can take the form of solids, liquids, sludges, or contained gases, and they are generated primarily by chemical production, manufacturing, and other industrial activities (the United States Environment Protection Agency (EPA), n.d.). For the agency, hazardous waste may cause damage during inadequate storage, transportation, treatment, or disposal operations.

The EPA mentions that improper hazardous-waste storage or disposal frequently contaminates surface water and groundwater supplies as harmful water pollution and can also be a source of dangerous land pollution. For instance, people living in homes built near old and abandoned waste disposal sites may be in a particularly vulnerable position. In an effort to remedy existing problems and to prevent future harm from hazardous wastes, governments closely regulate the practice of hazardous waste management (EPA, n.d.). The Waste Framework Directive lays down some basic waste management principles. It requires that waste be managed without endangering human health and harming the environment without risk to water, air, soil, plants or animals without causing a nuisance through noise or odors (European Commission, n.d.). The priority is to reduce the quantity and harmfulness of hazardous wastes.

Legal Deficiencies

Indeed, the management of hazardous waste presents several challenges that can hinder the effective implementation of the regulations. As you've pointed out, the legislation can sometimes be unclear and too vague in the absence of clear scientific criteria for the hazardousness of a waste. This can distort the distinction made by the law between the different categories of non-hazardous waste and waste, as waste, even household waste, can to a certain degree be dangerous and require a specific management procedure.

The issue of responsibility, particularly that of the producer, is another challenge. Any natural or legal person whose activity generates hazardous waste assumes responsibility for reducing quantitatively and qualitatively the harmful effects of the hazardous waste generated by the activity. However, this responsibility faces technical limitations, in particular the failure of the control and monitoring systems and the lack of financial means of the administration to ensure this control in a sufficiently efficient manner.

The legislation on waste does not really take a clear and precise legal definition of the notion of waste. There is also the insufficiency of the legal framework from the point of view of the sectoral approach; the control mechanisms are insufficient and the sanctions defined in this law are insufficient as well. The Law No. 28-00 is written in an incomplete legal style, which hinders the application of this law, and makes it imprecise. Furthermore, in the absence of hygiene measures, hazardous waste can directly harm human health if it is poorly managed. It should also be noted that the hazardous waste master plan is still in progress. This is due to constraints encountered during the initiation, development and implementation of this plan.

Although the legislation on hazardous waste provides for a multiplicity of provisions aimed at protecting nature against the harmful effects of waste, it suffers from several shortcomings that need to be reviewed. Regarding the legal and institutional shortcomings, it should be noted that hazardous waste has not been defined by the legislator, as the definition in the text is quite broad.

It should be noted that there is an absence of damage addressing in Law 28-00, which provides for sanctions in cases where the law is transgressed without taking into consideration the results of these acts. Therefore, regardless of the damage caused, the perpetrators of the offenses will have only one penalty to face. There is also a scarcity of prosecution, despite the repressive aspect, it can be seen on the ground that the competent authorities rarely prosecute companies and factories that pollute the environment with their toxic waste.

Lack of specific rules on the criminal liability of legal persons: Law No. 28-00 does not contain any provisions on the managers of legal entities. It should be noted that there is an overlap of competencies; Article 62 of Law No. 28-00 states in substance that an offense can be detected several times by the intervention of several agents, such a situation can lead to an encroachment of competence.

The penalties provided are not excessive enough as they range from 200 to 2,000,000 MAD and imprisonment from one month to three years. A scattering of the provisions governing waste, such as deficiency, is provided for by several legal texts, which leads to a lack of knowledge on the part of the agents authorized to carry out control activities. Moreover, the implementation of the plans is foreseen for ten years, which is a very long period and makes the implementation of these plans more delicate.

Reinforcement of Legal Measures Concerned with Sustainable Development

Indeed, addressing the various gaps in waste legislation, particularly in relation to hazardous waste, is crucial for sustainable development. Strengthening the legal framework for waste management and enhancing governance in the hazardous waste sector through access to information, transparency in service provision, monitoring, and evaluation are key steps in this process.

Replacing prison sentences for hazardous waste with rehabilitative sentences could be a more effective approach. Legislation should clarify recycling arrangements and introduce awareness-raising measures to encourage the adoption of hazardous waste recycling arrangements. Countries like Sweden have made significant progress in waste recycling, becoming a zero-waste country. It would be beneficial for Morocco to present an annual balance sheet of hazardous waste generation and set a national recycling target of 80 percent for these types of waste each year.

Waste can be recovered through several processes: recycling, and electrical or thermal energy production. Setting up a waste prevention program over a period of two years would aim at reducing hazardous waste in this sector through design, use, and improvement of sorting at source during deconstruction and rehabilitation operations, in particular of excavated earth.

Regions should be responsible for establishing a regional plan for the prevention and management of hazardous waste through a regional plan for sustainable development of the territory. This plan for the organization of hazardous waste management would be an interesting lever to include the use of recycled products in the specifications of public construction sites. Thus, in public tenders, favoring recycled hazardous waste would introduce criteria for reducing CO2 emissions. Furthermore, it is preferable to provide a text allowing the integration of environmental criteria in public purchases, allowing the introduction of clauses relating to products or processes, or selection criteria in public contracts. Public procurement should also be seized to promote eco-design and recycling, especially in the hazardous waste sector.

In this sense, it is important to financially support the development of companies' industrial tools to incorporate recycled hazardous waste and introduce a minimum rate of recycled waste in the specifications of public procurement. Additionally, encouraging research, development, and innovation to promote other ways of recovering hazardous waste is also a must.

,In addition, the State should provide support for the treatment and recovery of hazardous waste through a newly established national research agency. This agency would coordinate research projects in partnership with companies and channel funding to innovative industrial solutions. A dedicated Waste Fund should also be created to back territorial prevention and management initiatives and to finance investments in sorting, recycling, biological recovery (e.g., composting and anaerobic digestion), and energy recovery.

In the same line, a revision of the legislation on hazardous waste seems necessary to ensure better protection. In this sense, it is necessary to codify all the texts related to the environment, a codification that is strongly recommended as a solution to the complexity and dispersion of environmental texts.

The evolution of waste legislation reflects a shift in perspective where waste is no longer seen merely as a financial constraint or loss, but as a potential source of profit. This is evident in the growth of eco-industries in the recovery and reuse sectors over the last twenty years, and the widespread development of recycling and recovery programs in most companies and administrations.

To improve the current legislation, several principles could be introduced. These could take the form of provisional quantified objectives aimed at reducing the tonnage of hazardous waste, increasing recycling, and providing for future measures concerning the taxation of disposal facilities and that of the public waste management service.

The implementation of an efficient, economic, and organizational regulatory framework could aim at improving the treatment of certain hazardous waste and diffusing household waste, strengthening the means of energy recovery of the residual part of the waste, and improving territorial planning tools.

The legislator should also clarify the definitions of basic concepts such as waste, recovery, and disposal, strengthen waste prevention measures, introduce a life-cycle approach to products and materials and not just the waste phase, and focus on reducing the environmental and health impacts of waste generation and management, thereby enhancing the economic value of waste. In addition, waste recovery and the use of recovery materials should be encouraged in order to preserve natural resources.

One of the contributions that would be innovative is that the legislation should provide that producers and holders of hazardous waste must respect a hierarchy of treatment methods in the following order: preparation for reuse; recycling; recovery, including energy recovery; and disposal. Thus, operators of hazardous waste facilities should organize the management of their waste in accordance with this order of priorities, and territorial authorities may oblige them to do so.

It is appropriate to provide for new legal definitions and concepts that should be reformed by making innovative distinctions through new definitions, as the legal status of waste and its maintenance has become a real source of questioning. The definition of waste should be modified, qualifying it as any substance or object, or more generally any movable good, which the holder discards or intends or is obliged to discard.

It's important to note that the so-called liability mechanism is being revised into an extended liability, marking a significant trend in the ever-expanding law of hazardous waste. This trend sees an increase in the responsibility for products that generate waste, affecting the nature of the activities involved.

Extended producer responsibility could require producers, importers, and distributors of these products, or the elements and materials used in their production, to contribute to the management of hazardous waste derived from them. This, in turn, serves as an incentive to adopt eco-design for products.

This obligation could be fulfilled by establishing individual collection and treatment systems, subject to approval. Alternatively, state-approved eco-organizations could be set up, to which financial contributions would be made. This approach not only ensures the proper disposal of hazardous waste but also encourages the use of environmentally friendly product designs.

Conclusion

The waste processing and recovery sector indeed plays a significant role in the Moroccan economy, contributing to a substantial turnover and providing numerous jobs. The collection, sorting, and recycling of hazardous waste, which is the focus of this article, represent only a fraction of the Moroccan industry estimated to be worth around ten billion. It's suggested that companies' profits should be reinvested into recycling efforts. Public funding for these operations could be partially provided by public services dedicated to the management of hazardous waste. However, the introduction of eco-taxes in Morocco presents an interesting proposition. These taxes could potentially finance the collection and treatment of this waste, thereby ensuring a sustainable and environmentally friendly waste management system. This approach could also encourage Moroccan companies to adopt more eco-friendly practices to reduce their tax liabilities. Finally, we believe that hazardous waste management in Morocco is a complex issue that requires careful consideration and balanced decision-making.

Note

1 Dahir n°1-06-153of30 chaoual1427 (22 November 2006), OfficialBulletinn°5480 of 7 December 2006

References

Cirelli, C., & Maccaglia, F. (2019). *Territoires des déchets* [Territories of waste]. Presses Universitaires François-Rabelais.

Dahchour, A., & Hajjaji, S. E. (2020). Management of solid waste in Morocco. In A. Negm & N. Shareef (Eds.), *Waste management in MENA regions* (pp. 13–33). Springer Water. Springer. https://doi.org/10.1007/978-3-030-18350-9_2

Dahir n° 1-06-153 du 30 chaoual 1427 (2006, novembre 22), Bulletin officiel n° 5480 du 7 décembre 2006.

Décret n°2-14- 85 du 20 janvier 2015 relatif à la gestion des déchets dangereux.

European Commission (n.d.). *Waste Framework Directive*. Retrieved from https://environment.ec.europa.eu/topics/waste-and-recycling/waste-framework-directive_en

Hansen, C. (2023, October 23). *Waste management in Morocco*. Retrieved from https://www.ecomena.org/waste-management-morocco/

United States Environmet Protection Agency. (n.d.). *Learn the basics of hazardous waste*. Retrieved from https://www.epa.gov/hw/learn-basics-hazardous-waste

World Bank (n.d.). *Morocco: Municipal solid waste management program*. Retrieved from https://www.worldbank.org/content/dam/Worldbank/document/Climate/Climate-Finance-Projects-briefs/Morocco-Municipal-Solid-Waste-Management.pdf

6

The Protection of Water Rights in the Face of Human Security

Morocco as a Case

Rabiaa Bourhim

Morocco is a country with a semi-arid climate and is classified among the countries with high water stress. Water has always been a priority. For decades, Morocco has been committed to the management of its water resources by building large hydraulic infrastructures to meet its domestic, industrial, and agricultural needs. The discourse of public authorities and certain concrete actions demonstrate that there is a clear awareness of the value of water resources, accompanied by a political will to organize and control their management.

Being aware of the importance and topicality of the subject of water management in Morocco, and especially of the rational management that should take into account the impacts of climate change on natural resources and agriculture, Morocco has implemented a water policy and has taken legislative and institutional measures to better manage the resource and its distribution to the population. The current context of scarcity and climate change requires a rethinking of the governance systems for this essential resource.[1]

During the analysis of the Moroccan legal system linked to the management and exploitation of water, namely the Constitution, the laws on water, and even Islamic doctrine and endogenous standards, we discovered the existence of several legal orders. This leads to questioning the nature of their relationship: is it a competition, a rivalry?

The 2011 Constitution

Morocco, a signatory to a large number of multilateral instruments in the fields of environment and climate change since the early 1970s, has already made significant efforts to protect the environment, promote sustainable development, and create the essential framework conditions for the implementation of its national environmental policy. This environmental policy is based on the 2011 Constitution, the National Charter for the Environment and Sustainable Development, the National Strategy for Sustainable Development, and a range of legal and regulatory texts relating to the management and protection of natural resources such as water.

In relation to the environment and sustainable development, Morocco has undergone a transition in its constitutional, legal, and institutional framework. To that end, the King of Morocco, Mohammed VI, stated in his Royal Address to the Nation on the occasion of the nineteenth anniversary of the Sovereign's accession to the throne in 2018: "Our constant concern to improve the general social situation and to face economic challenges is matched only by Our active commitment to preserve and enhance our country's strategic resources."[2] In this regard, first and foremost, water plays a key role in the dynamics of development and the preservation of stability.

In this sense, and on the occasion of Throne Day, July 30, 2009, the King of Morocco declares that "Morocco, which, like all developing countries, faces major and pressing challenges in terms of development, is fully aware of the need to preserve the environment and to meet ecological imperatives." In light of these requirements and commitments, (. . .) it is necessary to pursue a policy of gradual and comprehensive upgrading, both economically and in terms of awareness, with the help of regional and international partners. In this spirit, Morocco adopted a new constitution in 2011, which is the ultimate support for sustainable development and environmental protection. This constitution places the environment among the fundamental rights of citizens by stating in Article 19 that "men and women enjoy equal rights and freedoms of a civil, political, economic, social, cultural, and environmental nature." Article 31 of the constitution states that "public institutions and local authorities shall work to mobilize all available means to facilitate equal access for citizens to the conditions that allow them to enjoy the rights (. . .) to access water, a healthy environment, and sustainable development."[3]

Indeed, the inclusion of environmental protection, the promotion of sustainable development, and natural resources in the Constitution of 2011 is indicative of an acceleration of ecological awareness and evidence of the growing importance of these areas for the evolution of Moroccan society. Thus, the country has embarked on a process of reforms affecting various areas, including the environment, water, sustainable development, and climate. In addition, the right to a healthy environment, now constitutionalized, has been consolidated by the Framework Law instituting the National Charter for the Environment and Sustainable Development, which states in its Article 3 that "every citizen has the right: to live and evolve in a healthy and quality environment that promotes the preservation of health, cultural development, and sustainable use of the heritage and resources available therein; to have access to reliable and relevant environmental information; and to participate in the process of making decisions that may have an impact on the environment." Moreover, the constitutional recognition of sustainable development as a human right clearly demonstrates the interest in this concept in Morocco.

In this sense, Article 71 of the Constitution provides that the rules relating to environmental management, the protection of natural resources, and sustainable development, and the regime of water, forests, and fisheries fall within the domain of the law, in addition to the matters expressly assigned to it by other articles of the Constitution. It is true that sustainable development entered the Moroccan legal arsenal in 2003 through Law 11-03 on the protection and development of the environment. In 2011, the Constitution recognized the right to sustainable development and referred to sustainable development on several occasions, including through the establishment of the Economic, Social and Environmental Council. In 2014, the National Charter for the Environment and Sustainable Development was adopted. Morocco thus displays its choice, unequivocally, for sustainable development. However, the charter did not introduce essential elements related to the management and protection of the environment and sustainability; the most striking is gender and the important role played by women, especially in the rural world.

Water Laws

In Morocco, available water resources are considered among the lowest in the world when compared to the country's population, knowing that this population is expected to increase by 300,000 inhabitants per year to reach 41.9 million in 2030[4]. The consumption needs will be increasingly high, and the resources less and less available, which will make the question of water a vital issue for the country. The available water resources reach 22 billion cubic metres per year and are distributed between 18 billion of surface water[5] (lakes, rivers, etc.) and 4 billion of groundwater. With an annual water availability of 700 cubic metres per capita[6], Morocco is already a country with low water resources. These resources are subject to several anthropogenic pressures (overexploitation of groundwater, pollution and discharge of waste and untreated wastewater, climate change and variability), which lead to a decrease in the quantity and quality of water resources and impact on the socio-ecological systems of the country.

Other factors, such as inadequate water supply, sanitation, and hygiene practices, have negative health effects, such as diarrhea and malnutrition. These factors cause losses estimated at 11.7 billion MAD, or 1.26 percent of GDP[7]. Specifically, the overexploitation of groundwater and the discharge of untreated industrial water pose serious problems. In 1995, Morocco adopted Law No. 10-95 on Water, which marks a fundamental step in the state's awareness of the importance of strengthening the legal framework in this area. Indeed, this law, which outlines the legal framework of national water policy, not only unified existing texts but also introduced many novelties, such as the establishment of national legal bases for forecasting the evolution of resources and needs, the rationalization and efficient decentralized management, the reduction of inequalities, and the development of inter-regional solidarity. In addition, this law aims to achieve the following objectives: rational and optimal use of water resources, both at the regional and national levels, taking into account the orders of priority set by the national water plan; and the management, protection, and conservation of water resources within the framework of the hydraulic basin, which is an important innovation allowing the design and implementation of decentralized water management[8].

In 2005, the law was supplemented by Enforcement Decree No. 2-04-553 on discharges, flows, and discharges that result in direct or indirect deposits in surface or groundwater. This text opens, in turn, important perspectives as regards the establishment of standards intended to ensure the effective protection of existing resources. In the same sense, Law No. 10-95 was completed by Law No. 36-15 of 2016. The latter brings many novelties, such as the strengthening of the legal framework and encouraging the recovery of rainwater, wastewater, and seawater. The text also integrates the climate imperative in order to protect water resources against the impacts of extreme events induced by climate change. It is based on a set of principles, including general ownership of water; the generalized right of access to water; the right to a healthy environment; water management through inclusive governance based on the participatory approach; protection of natural environments, development of sustainable management; and the gender approach. In addition, Law n° 36-15 introduces a novelty concerning the creation of consultative councils for each water basin in Morocco to ensure optimal management of resources.[9]

In general, the laws adopted in the water sector have led to the existence of multiple institutions and bodies whose prerogatives often overlap, which is an obstacle to good water governance in Morocco. This multitude of actors is often doomed to failure due to a lack of clarity in the roles and responsibilities of each and the lack of coordination between these actors. In this vein, the Moroccan Economic, Social, and Environmental Council stated in its 2014 report entitled "Governance through Integrated Water Resources Management in Morocco: A Fundamental Lever for Sustainable Development" that "due to the various actors' multiplicity and situations ranging from multi-coverage to lack of coverage, the overall scheme of interoperability becomes extremely complex, generating major risks, and therefore inefficient" . The overlap and dispersion in the regulatory texts between several ministerial departments and the plurality of actors constitute the major hindrance to good governance of water management, as more than twenty actors are responsible for water management in Morocco:[10]

- Law 10-95 (August 16, 1995) on water that evokes principles such as "polluter pays" and "levy payer"
- Law 36-15 (August 10, 2016) revises and supplements Law 10.95 on water and sets the basis for an integrated, decentralized, and participatory management of water resources
- The Higher Council for Water and Climate
- The Hydraulic Basin Agencies
- The Councils of Hydraulic Basins
- The National Office of Electricity and Drinking Water
- The Prefectural and Provincial Water Commissions
- The National Agency for Water and Forests
- The National Council for the Environment
- The Secretariat of State for Water and the Environment
- The Interministerial Water Commission
- The National Program for Drinking Water Supply and Irrigation 2020–2027
- The National Water Plan for the period 2020–2050

- The Integrated Water Resources Development Master Plan
- The Economic, Social, and Environmental Council
- The National Program of Water Saving in Irrigation
- The Desalination of Sea Water and the Demineralization of Brackish Water
- The National Plan for Liquid Sanitation
- The National Program for Solid Household Waste
- The Moroccan Coalition for Water
- The National Mutualised Sanitation Program
- In the old title, the Haut-commissariat aux Eaux et Forêts et à la Désertification.

However, despite this entire legal and institutional arsenal, Morocco, according to the report published by the World Bank in 2017, remains as a country classified below the "extreme water stress" level defined by the United Nations and below the "water poverty level" with 645 cubic metres of water per person per year. The right to water is defined by Resolution 64/292 of the United Nations General Assembly on July 28, 2010, and subsequent resolutions, entitled "The human rights to drinking water and sanitation," adopted by the United Nations Human Rights Council. It is also enshrined in Article 31 of our 2011 Constitution.[11]

Indeed, despite the ambitious dam policy adopted by Morocco, several reports state that "Morocco is expected to face extreme water scarcity by 2050. Considering the severe and negative impacts of future climate change, the country could be pushed well below the extreme water scarcity threshold in the coming decades."[12]

The Moroccan Economic, Social, and Environmental Council is sounding the alarm and calling for urgent measures to be taken.[13] These measures include designing and implementing a communication strategy to raise awareness among all users about the vital importance of adopting eco-responsible behavior regarding water. Additionally, the Council advocates breaking with the practice of watering public green spaces, sports facilities, and recreational parks with drinking water by systematically reusing wastewater. It also emphasizes the need to systematize the audit and detection of leaks to elevate the performance of water supply and distribution in cities to international standards. Furthermore, the Council calls for systematizing the study of impact and water efficiency for investment projects, particularly in agriculture, industry, and tourism, and introducing recourse to arbitration by the Head of Government in this area. Initiating a thorough reform of national and local water pricing and liquid sanitation and wastewater treatment services, and communicating the results, is also highlighted. Establishing a national water accounting framework that reflects the real costs of water by watershed, ensuring regional and social solidarity through better targeting of public subsidies to the sector, and optimizing the regions' self-financing capacities are crucial steps. Lastly, the Council stresses the importance of activating the coordination and consultation bodies at the national and local levels to make arbitration decisions, not only on conflicts but also on the various criteria arising from an investment and sustainable development policy. Strategic measures for better allocation of investments in the water sector are also included.

Patrimonialization of the Commons in Morocco

The situation of water stress, scarcity, and drought that Morocco has experienced this year calls into question all the processes and public policies adopted and calls for the implementation of new endogenous dimensions and resilient paradigms capable of managing and protecting this vital resource, such as the concepts of heritage and commons.

The endogenous practices of water management constitute an international intangible heritage of humanity, a key element in the preservation of the social fabric and natural resources, and a factor of solidarity and social cohesion to reduce the migration of populations and ensure the stability of peoples (Bourhim, 2022). In this sense, the coordinator of the NGO "Africa Civil Society Network for Water," Ms. Sareen Malik, stressed that customary water management is not specific to Morocco but to Africa as a whole, noting that any water management must combine the modern and traditional and be based on governance.[14]

In Morocco, the endogenous management practices present a normative reserve, a national heritage, and a civilizational genius. The khettaras[15] (Figure 6.1), Tighiria, and Tassa (Figure 6.2) present inescapable, resilient endogenous models[16] that enrich biodiversity, establish water sustainability, ensure the transparency of endogenous water management practices and equitable sharing, reinforce the sense of responsibility of each user, and avoid the overconsumption of natural resources. Indeed, the notion of commons and patrimony can trigger a constructive dialogue between positivism and endogeneity and bring out alternative development modes based on a new systemic approach to water in a semi-arid country. "Water is a common environmental good par excellence."

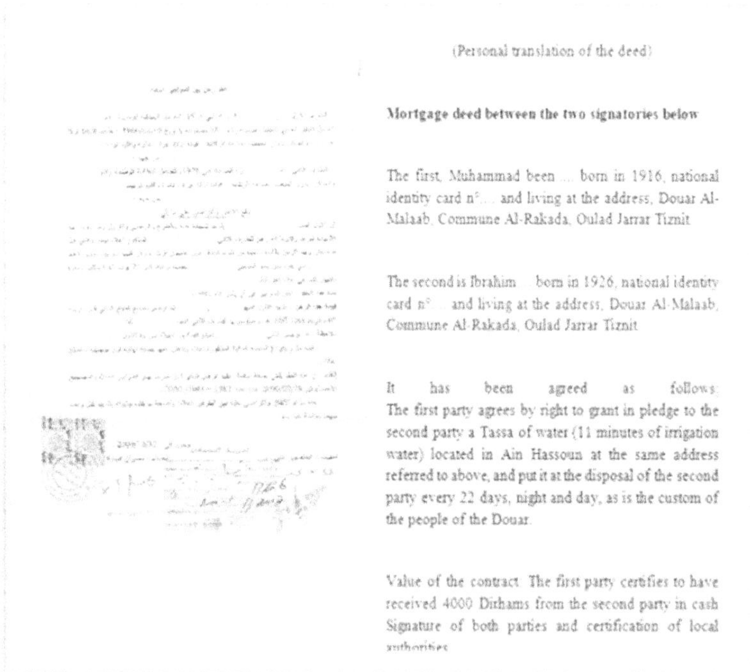

(Personal translation of the deed)

Mortgage deed between the two signatories below

The first, Muhammad been .... born in 1916, national identity card n°.... and living at the address, Douar Al-Malaab, Commune Al-Rakada, Oulad Jarrar Tiznit

The second is Ibrahim ... born in 1926, national identity card n°.... and living at the address, Douar Al-Malaab, Commune Al-Rakada, Oulad Jarrar Tiznit

It has been agreed as follows:
The first party agrees by right to grant in pledge to the second party a Tassa of water (11 minutes of irrigation water) located in Ain Hassoun at the same address referred to above, and put it at the disposal of the second party every 22 days, night and day, as is the custom of the people of the Douar.

Value of the contract: The first party certifies to have received 4000 Dirhams from the second party in cash
Signature of both parties and certification of local authorities

Figure 6.1 Endogenous Mortgage Act linked to the Right to Use Irrigation Water (Tighira or Tassa). Source: Rabiaa Bourhim.

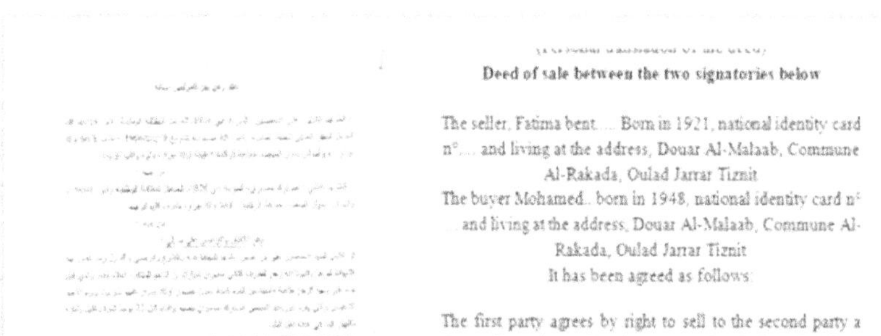

(Personal translation of the deed)

Deed of sale between the two signatories below

The seller, Fatima bent .... Born in 1921, national identity card n°.... and living at the address, Douar Al-Malaab, Commune Al-Rakada, Oulad Jarrar Tiznit
The buyer Mohamed.. born in 1948, national identity card n° .... and living at the address, Douar Al-Malaab, Commune Al-Rakada, Oulad Jarrar Tiznit
It has been agreed as follows:

The first party agrees by right to sell to the second party a Tassa of water located in Ain Hassoun that she inherited from her father at the same address mentioned above, and put it at the disposal of the second party every 22 days, night and day, as is the custom of the people of the Douar
Value of the contract: The first party certifies to have received 6000 Dirhams from the second party in cash

Signature of both parties and certification of local authorities

16.12.2015

Figure 6.2 Endogenous deed of sale linked to the Right to Use Irrigation Water (Tighira or Tassa). Source: Rabiaa Bourhim.

Indispensable to the survival of not only humans but also ecological systems, it meets the most basic needs and must be shared between communities that, although inhabit the same blue planet, do not have the same access to this resource. As a result, many authors have argued that the wars of the twenty-first century will be fought over "water conflicts" (Petitjean, 2016; Lasserre, 2009), and that water heritage (which includes traditional know-how, customary law, and traditional rites and uses of water) could help provide more sustainable legitimate and post-modern solutions[17]. The customary norms and know-how can serve as a legal basis for all the legal rights of a community, based on its traditional knowledge, in terms of sharing and management, and the regulation of conflicts and disputes. The distinctiveness stems from the composition of a set of customs,[18], beliefs, traditional knowledge, and values accepted as obligatory rules of conduct and ways of life in communities that are integral parts of socioeconomic systems (Bourhim, 2021).

In Morocco, custom has played a considerable role as a normative source that can inspire positive law by proposing solutions drawn from the common historical legal fund of Moroccans. It constitutes a reservoir of legal solutions that are adequate and appropriate to Moroccan mentalities without undermining national and religious principles. Indeed, the concept of "custom" has more than one meaning. It can be considered "a social fact," which is not singular and resists more to the effect of erosion. It is manifested by its regularity, repetition, and, above all, generality. Thus, a social fact remains an ephemeral social phenomenon if these conditions are not met. Montagne believes, in his article devoted to the study of the legal system of southern Morocco, that "customary law is the prerogative of the Siba tribes and tends to disappear with the penetration of the Makhzen, which substitutes itself for the customary rules of application of the religious rule." (Montagne, 1924)

As a result, "custom" can also be defined as "a norm of objective law based on a popular tradition that lends a constant practice a legally binding character; a true rule of law but of non-state origin" (Bourel, 2014). To summarize, as a reference, customary rules continue to be the most relevant to local realities. However, these rules would suffer from weak legitimacy once considered from the perspective of positive law. The goal is not the codification of local knowledge, but rather the strengthening of the consensual decision-making capacities of local communities, which constitutes the assurance of optimal regulation on the territory. This knowledge, which constitutes local communities' civilizational genius, must inspire legislators in the creation of law. The latter, because it applies to man, must therefore emanate from him.

To this end, decentralization is the most appropriate institutional structure for sustainable management and an imperative for conservative environmental management. This decentralization stems from the principle of subsidiary, according to which all decisions must be taken at the lowest possible level. Thus, the involvement and commitment of key actors around common challenges and societal projects serve as the catalyst for a negotiated approach rather than a strictly planning approach. It is also a heritage approach that mobilizes the local population around a collective issue. The objective is to protect and preserve a common heritage, ensuring the security and future of the social group. Such an approach goes towards a new logic based on patrimonialization as a relational form of social groups and space.

Elinor Ostrom confirms that communities have an important role to play in preserving heritage, although the modes of preservation may vary. This heritage approach to water management can achieve objectives that avoid overexploitation, conserve the resource, and avoid conflicts (Simon, 2021). The intangible heritage approach developed by Ostrom has expanded the scope of this study and the understanding of concepts such as water heritage and has helped us to further explore the importance of communities (Simon, 2021). The network of khettaras, the Tighira system, and the Inflassen ensure the distribution of surface water in rural areas still in use and serve to share irrigation water while following well-established rules of allocation and being managed by the local community.

Conclusion

In conclusion, the severe water stress in Morocco is compelling decision-makers to implement innovative, legally sound, and contextually appropriate water management strategies. The patrimonialization of the commons facilitates an interaction between the human, ecological, economic, and social dimensions. These new management methods aim to consolidate these dimensions to achieve the country's sustainable development objectives. The insights gained from these approaches allow us to draw significant conclusions that could enhance our understanding of not only the notions of heritage and the commons but also the intricate connections between them. Indeed, water heritage delineates the coexistence between humans, the community, and the biosphere with this natural resource that they utilize and safeguard. The endogenous knowledge and infrastructures that contribute to water management exist fundamentally because water is recognized as a divine gift, a resource before which all living beings are equal.

Notes

1 Houria Tazi Sadeq La Revue De L'institut Veolia - Facts Reports N° 22 - Water, waste, energy: What future for essential services in Africa? 2021
2 https://www.maroc.ma/fr/discours-royaux/sm-le-roi-adresse-un-discours-la-nation-loccasion-de-la-fete-du-trone-0
3 https://faolex.fao.org/docs/pdf/Mor185348.pdf
4 Haut-commissariat au Plan, "Prospective Maroc 2030: Quelle démographie?" Study published on August 8, 2011, http://www.hcp.ma/downloads/Maroc-2030_t11885.html
5 Water Department of the Ministry of Energy, Mines, Water and Environment, "Water Policy in Morocco," Presentation, September 2012.
6 https://www.kas.de/fr/web/marokko/laenderberichte/detail/-/content/la-gestion-de-l-eau-au-maroc1
7 https://leseco.ma/business/degradation-de-l-environnement-un-cout-evalue-a-3-5-du-pib.html
8 Introduction of the law n° 10-95 on Water
9 Introduction of the law n° 10-95 on Water
10 https://ladepeche24.com/etat-durgence-hydrique-decrete-au-maroc-pourquoi-arriver-au-stress-hydrique-extreme/
11 https://ladepeche24.com/etat-durgence-hydrique-decrete-au-maroc-pourquoi-arriver-au-stress-hydrique-extreme/
12 « Gestion de la Rareté de l'Eau en Milieu Urbain au Maroc», Rapport de la banque mondiale, 2017 in: https://documents1.worldbank.org/curated/en/488091516133312338/pdf/summary-report.pdf
13 http://www.cese.ma/media/2020/11/Alerte-CESE-VF-1.pdf
14 https://www.maroc.ma/fr/actualites/m-toufiq-les-pratiques-traditionnelles-de-gestion-de-leau-continuent-dinteresser-et-de
15 The Khettaras represent a real intangible heritage and an ingenious ancestral know-how of drainage of underground water played through history a primordial role in the development of the cities, Unlike the canals, which tend to be invaded by vegetation, and are very sensitive to the bad weather and prone to silting, the

khettaras make it possible to bring a healthy water with a minimum of evaporation. These underground galleries were used to drain water from the mountains from the lower Atlas Mountains https://www.agrimaroc.ma/les-khettara-un-systeme-traditionnel-ingenieux-de-drainage-des-eaux-en-danger-de-disparition/, accessed on 22/12/2022.

16 Resilience is not, therefore, a synonym for resistance to change, but rather the ability to actively transform and adapt, based on scientific and technical progress, communication, negotiation and the use of local knowledge (Dentz & Bailli, 2005). Resilience is the capacity of a system to absorb all disturbances and to reorganize itself while making changes in such a way as to keep essentially the same functions, structures, identity and feedback (Holling, 1973)

17 In his 1965 dissertation on groundwater management in Southern California, Ostrom empirically observed a wide variety of commons systems. The latter generally combine self-management, customary rules, market mechanisms and state regulation. In this view, Ostrom notes that "Effective and sustainable management of a common pool natural resource is also more likely to be accomplished when the boundaries of the resource are easily identified, when its changes in state can be monitored at relatively low cost, when the rate of change in socio-economic, technological and use conditions remains stable, when communities maintain frequent social interactions with each other and thus foster trust in each other (thus increasing social capital), when outsiders can be relatively easily prevented from accessing the resource (thus preventing its widespread use), and when violations of the rules are monitored and punished (Le Roy, 2012)." Thus, Ostrom demonstrates that user communities often manage to invent effective and robust management systems on their own.

Custom is the product of the group's consensual process, and for custom to exist, the group must be alive and independent. Reinforcing this assertion, Crozier considers the organization as "a political and cultural construct," a means by which social actors regulate their interactions in order to obtain the minimum cooperation necessary to achieve collective goals.

The Khettaras represent a real intangible heritage and an ingenious ancestral know-how of drainage of underground water played through history a primordial role in the development of the cities, Unlike the canals, which tend to be invaded by vegetation, and are very sensitive to the bad weather and prone to silting, the khettaras make it possible to bring a healthy water with a minimum of evaporation. These underground galleries were used to drain water from the mountains from the lower Atlas Mountains https://www.agrimaroc.ma/les-khettara-un-systeme-traditionnel-ingenieux-de-drainage-des-eaux-en-danger-de-disparition/, accessed on 22/12/2022.

18 Custom is the product of the group's consensual process, and for custom to exist, the group must be alive and independent. Reinforcing this assertion, Crozier considers the organization as "a political and cultural construct, a means by which social actors regulate their interactions in order to obtain the minimum cooperation necessary to achieve collective goals."

References

Bourel, É. (2014). Ott-Duclaux-Monteil Cécile, 2013, Exploitation forestière et droits des populations en Afrique centrale. *Journal des africanistes*, *8*(4–1), 301–303.

Bourhim, R. (2021). *De la durabilité à la coviabilité socioécologique: vers l'invention d'un droit post-moderne dans l'espace collectif marocain* [From sustainability to socioecological cohabitation: Towards the invention of a post-modern law in the Moroccan collective space.] *REMALD*.

Bourhim, R. (2022). Resolution of user rights related conflicts in collective rangelands through negotiation: The case of Morocco». In *The climate-conflict-displacement nexus from a human security perspective* (pp. 353–371). Springer.

Dentz, E., & Bailli, G. (2005). *La résilience dans l'entreprise: après le choc, disparaître ou rebondir* [Resilience in business: after the shock, disappear or bounce back.] Hermès Science.

Gestion de la Rareté de l'Eau en Milieu Urbain au Maroc, Rapport de la banque mondiale, 2017 disponible sur le site https://documents1.worldbank.org/curated/en/488091516133312338/pdf/summary-report.pdf, consulté le December 20, 2022.

Haut-Commissariat au Plan, «Prospective Maroc 2030: Quelle démographie? étude publiée le 08 août 2011, http://www.hcp.ma/downloads/Maroc-2030_t11885.html

Holling, C. S. (1973). Resilience and stability of ecological systems». *Annual Review of Ecology and Systematics*, *4*(1), 1–23.

Houria Tazi Sadeq LA REVUE DE L'INSTITUT VEOLIA - FACTS REPORTS N° 22 - Eau, déchets, énergie: Quel avenir pour les services essentiels en Afrique? 2021

http://www.cese.ma/media/2020/11/Alerte-CESE-VF-1.pdf

https://ladepeche24.com/etat-durgence-hydrique-decrete-au-maroc-pourquoi-arriver-au-stress-hydrique-extreme/

https://www.kas.de/fr/web/marokko/laenderberichte/detail/-/content/la-gestion-de-l-eau-au-maroc1

Lasserre, F. (2009). Les guerres de l'eau: L'eau au cœur des conflits du XXIe siècle [Water Wars: Water at the Heart of 21st Century Conflicts]. Delavilla. ISBN 978-2917986028.

Le Roy, A. (2012). Des communs sans tragédie: Elinor Oström vs. Garrett Hardin [Commons without tragedy: Elinor Ostrom vs. Garrett Hardin]. *EcoRev'*, (1), 24–27.

Montagne, R. (1924). *Le régime juridique des tribus du sud marocain* [The legal regime of the tribes of southern Morocco]. E. Larose.

Petitjean, O. (2016, June 30). Les guerres de l'eau auront-elles lieu ? [Will water wars take place?] Ritimo / Partage des eaux. https://www.ritimo.org/Les-guerres-de-l-eau-auront-elles-lieu

Simon, S. (2021). Réintroduire le patrimoine de l'eau parmi les communs. Pour un développement alternatif et durable au Maroc [Re-establishing water heritage among the commons. For an alternative and sustainable development in Morocco]. *In Situ. Au regard des sciences sociales.* https://doi.org/10.4000/insituarss.599

7

The Right to Healthy Environment in Morocco Between International Conventions and National Legislation

Fatima Zohra Alaoui Mahrez and Hind Hassine

The right to a healthy, balanced, and clean environment is among the most important demands and purposes on which the wheel of modernity and development should be based, especially due to the severe environmental pollution that the globe is suffering from. This pollution has resulted in many problems and chronic diseases such as asthma and allergies, among others. Moreover, it has led to the emergence of global warming and turbulent climate change, which portend serious environmental disasters that will threaten not only the economies of states but also the very existence of humanity (Chalabi, 2015).

For many decades, the topic of the environment has been one of the most important and controversial issues all over the world. Both national and international laws have tried to set rules to protect its resources and preserve them from damage as they are a common heritage for the entire humanity. In this sense, the problematic nature of the environment intersects with diverse economic, political, social, cultural, and humanitarian harms.

According to Article 3 of Law 11.03, pertaining to The Protection and Enhancement of the Environment 2003, the environment is defined as "all the natural elements and human settlements as well as the economic, social, and cultural factors that further the existence and development of living organisms and of human activities." Its protection is achieved by "the preservation and improvement of the constituents of the environment, the prevention of their degradation and their pollution or the reduction of the said pollution" (Law 11.03, 2003, art 3, §I).

However, since Morocco has suffered from the harmful effects of industrial and technological development, it has seriously acted to harness all its efforts to preserve and protect its environment. Hence, it constitutionalized the right to live in a healthy environment in the Kingdom's Constitution of 2011, in addition to enacting laws and creating institutions that aim to achieve the same goal.

In this regard, this chapter tries to shed light on the role of the Kingdom of Morocco in the preservation of the right to a healthy and protected environment. Thus, this paper tries to answer the following questions:

- What are the main laws and procedures enacted and taken by Morocco to protect the right to a healthy environment?
- What is the role of international conventions, royal speeches, and Moroccan institutions to foster this right?

To answer these questions, this chapter is divided into three main parts. The first part intends to discuss the international conventions and treaties signed by Morocco in the field of environmental protection, as well as the royal speeches that highlight this protection. The second part focuses on Morocco's national legislation and the laws that govern and frame this field. Finally, the third part mentions the national institutions concerned with environmental protection and the main programs and projects conducted by Morocco in this field.

International Conventions and Royal Speeches for Environmental Security

Indeed, Morocco has made significant strides in its economic development, with key socio-economic sectors such as agriculture, industry, fishing, urban development, tourism, and infrastructure experiencing substantial growth. However, this progress has not been without environmental consequences. In response to these challenges, Morocco has taken on a global responsibility to protect the environment, implementing a series of measures to achieve this goal. The country has signed numerous international conventions that affirm the right to a healthy environment. His Majesty Mohamed VI, King of Morocco, representing the highest authority in the country, has emphasized the importance of environmental protection and preservation on numerous national and international occasions, reinforcing its significance in Morocco's policies. This commitment to environmental stewardship underscores Morocco's dedication to sustainable development and responsible growth.

International Protection of the Right to Healthy Environment

The phenomenon of environmental protection has received great attention at different levels. This attention has contributed in creating, directing and connecting international cooperation to address ecological problems. Amongst these cooperations, light can be shed on the international conventions and treaties signed by states to preserve the ecosystem. Morocco has been one of the most active agents in this field, and has shown its commitment by signing and ratifying numerous conventions in this regard. Figure 7.1[1] introduces some of these conventions in different fields such as maritime regulation, climate change, plants and animal protection, heritage protection, waste and hazardous chemicals

However, this section attempts to discuss some international conventions signed and ratified by Morocco in the field of environmental protection.

Figure 7.1 Main international conventions ratified by Morocco for Environment Protection. Source: Ministry of Energy Transition and Sustainable Development.

United Nations Framework Convention on Climate Change 1992

The United Nations Framework Convention on Climate Change (UNFCCC) was established to provide guidelines for nations in addressing climate change, one of the most pressing global issues. Adopted by many countries in 1992, it was signed during the International Conference on Environment and Development in Rio de Janeiro the same year. The convention came into effect in 1994 and has since grown to include 198 member states, including Morocco, which signed the convention in 1992 and ratified it in 1995.

The UNFCCC serves as a guide for its member states in combating climate change and is the first convention to address this issue. While it does not impose specific obligations on its parties, it outlines a set of principles and responsibilities that they should adhere to. Its primary objective is the "stabilization of greenhouse gas concentrations in the atmosphere at a level that would prevent dangerous anthropogenic interference with the climate system." It also aims to "ensure that food production is not threatened and enable economic development to proceed in a sustainable manner" (United Nations Framework Convention on Climate Change, 1992, p. 9). This underscores the convention's commitment to balancing environmental protection with economic growth.

Kyoto Protocol to the United Nations Framework Convention on Climate Change

Since the United Nations Framework Convention on Climate Change lacked specific and clear commitments and obligations for its parties, the Kyoto Protocol came to operationalize the convention by committing its parties, especially industrial and developed countries, to reduce greenhouse gases. This Protocol was adopted in 1997 and entered into force in 2005 due to its complex ratification process. The Protocol consists of twenty-eight articles that include the parties concerned, the pollutants harmful to the environment, the measures to be taken, the ways to implement international obligations individually or collectively, the desired relationship between developed and developing countries in this regard, and the methods of international follow-up to implement the agreement. Moreover, the convention binds only the developed countries and the countries that are undergoing the process of transition to a market economy, which are responsible for the increased polluting GHG emissions in the air. Consequently, Annex-A, in the aforementioned protocol, defines polluting gases and activities in energy, industry, agriculture, and transport, while Annex-B sets a list of countries concerned with reducing emissions and the percentage that they must reduce compared to the amount that was emitted in 1990. After many negotiations, the concerned parties agreed on several principles related to the objectives and procedures for implementation, follow-up, and evaluation. These principles include the need to reduce the concentration of greenhouse gases to levels that do not harm the ecosystem, promoting sustainable forms of agriculture in light of considerations related to climatic changes, and taking precautionary measures regarding emissions. Additionally, they emphasized the importance of carrying out scientific research related to new forms of energy and carbon dioxide sequestration technologies.

Indeed, Morocco ratified the Kyoto Protocol in 2002 and hosted the seventh Conference of the Parties (COP 7) in October 2001 in Marrakech, where it presented its first national communication. It's worth noting that Morocco has been very proactive in environmental protection and has demonstrated its commitment to addressing climate change on a global scale. Despite Morocco's greenhouse gas emissions being relatively small, the country has not shied away from taking practical steps to mitigate greenhouse gas emissions. This includes energy-saving projects, initiatives to control energy use, and the promotion of renewable energy sources (Climate Change Secretariat (UNFCCC), 2005, p. 65). This underscores Morocco's dedication to sustainable development and responsible growth.

The 1992 Convention on Biological Diversity

The international community adopted the Convention on Biological Diversity along with the United Nations Framework Convention at the Earth Summit in Reo De Janeiro in 1992 and entered into force in 1993. Morocco, in addition to other nation states, signed the convention in 1992 and ratified it on August 21, 1995, which confirms the importance it attaches to its living wealth and biological resources, including ecosystems and living species. The main objectives of the convention are:

> The conservation of biological diversity, the sustainable use of its components and the fair and equitable sharing of the benefits arising out of the utilization of

> genetic resources, including by appropriate access to genetic resources and by appropriate transfer of relevant technologies, taking into account all rights over those resources and to technologies, and by appropriate funding. (Convention on Biological Diversity, 1992, A.1, p. 3)

Furthermore, the Convention on Biological Diversity considers that the conservation of biodiversity is a common concern of humanity. Even if the constituents of biodiversity lie within national borders, biodiversity is essential to all life forms on Earth. Hence, collaborative work becomes a must to ensure its conservation. The Convention on Biological Diversity has adopted a key approach in its attempt to prevent biodiversity loss, which is the "ecosystem approach," seen as a strategy that supports the integrated management of soil, water, and living resources to encourage conservation and sustainable use. As a result, the Convention on Biological Diversity has bet on the preservation of genes, species, and ecosystems in their natural environment. Additionally, the Conference of the Parties is the supreme authority of the Convention that brings together representatives of all parties to the Convention, as well as observers, such as United Nations agencies, international organizations, and non-governmental organizations. Its function is defined in Article 23 of the Convention on Biological Diversity, which is to direct and supervise the entire implementation and development of the Convention. The Conference of the Parties reviews progress and develops action plans for further action. It can also propose amendments to the agreement and work in cooperation with other international treaties.

Basel Convention on Transboundary Hazardous Waste Control 1989

Indeed, the Basel Convention on the Control of Transboundary Movements of Hazardous Wastes and their Disposal, adopted on March 22, 1989, was a response to international demands to prevent the movement of hazardous wastes between countries. Specifically, it aimed to prevent the transfer of hazardous wastes from developed to developing countries and to address the movement of radioactive wastes. The agreement also seeks to reduce the amount of waste generated, to ensure environmentally sound management as far as possible, and to assist less developed countries in the environmentally sound management of the hazardous waste they generate. This underscores the global commitment to responsible waste management and environmental protection.

This agreement entered into force in 1995, in response to the annual global production of millions of tons of waste hazardous to human health and the environment, and the urgent need for international measures to deal with the transboundary movement of these wastes. In this regard, the main objectives of this convention are as follows:

> (i) the reduction of hazardous waste generation and the promotion of environmentally sound management of hazardous wastes, wherever the place of disposal; (ii) the restriction of transboundary movements of hazardous wastes except where it is perceived to be in accordance with the principles of environmentally sound management; (iii) and a regulatory system applying to cases where transboundary movements are permissible. (Basel Convention on the Control of Transboundary Movements of Hazardous Wastes and their Disposal, 1989, p. 5)

According to this agreement, the parties cannot prevent the import of hazardous and other wastes except after informing the other parties of this decision, and they cannot allow the export of hazardous wastes to the same parties that banned their import. The parties must prevent the export of toxic and other waste if the importing party does not express its consent in writing to this import, and prevent the transfer or disposal of hazardous and other waste if the disposal party does not express its consent in writing to this disposal. Additionally, they should inform the neighboring state party of the danger to health or the environment that the region may be exposed to as a result of any accident during the transportation or disposal of hazardous and other wastes (Basel Convention on the Control of Transboundary Movements of Hazardous Wastes and their Disposal, 1989). Furthermore, in December 1999, the Basel Protocol on Liability and Compensation for Damage Resulting from Transboundary Movements of Hazardous Wastes and Their Disposal was ratified. This protocol deals with losses that may result during transboundary movements of hazardous and other wastes, including illegal trafficking and methods of disposal.

Morocco, in turn, became involved in this agreement, as it signed it in 1989 and ratified it on December 28, 1995. Hence, the convention entered into force for Morocco on March 27, 1996. In this vein, Morocco has taken many measures to respect the obligation of this convention and achieve its aims. Morocco pledged to take a set of measures to implement the requirements of this agreement, including choosing the delegated ministry in charge of the environment as the main authority responsible for the Basel Convention. Additionally, Morocco committed to participating in conferences of the parties to the Basel Convention and organizing the African Conference on Hazardous Waste Management in 2001. An inventory of hazardous waste resulting from industrial and hospital activities was completed in 2000, followed by the completion of an inventory of polychlorinated biphenyls in 2002 and 2005. Furthermore, Morocco formulated a national plan for the management of hazardous waste and initiated a project to establish a national center for the disposal of special waste. The issuance of Law No. 28-00 related to waste management and disposal is also a significant step, with the liability and compensation protocol currently in progress (Ministère de La Transition Energétique et de Développement Durable, n.d.).

The Right to a Healthy Environment Through Royal Speeches

The topic of environmental protection has received special attention in the Royal speeches delivered by His Majesty King Mohammed VI in national and international summits and events. The speeches in question present an environmental discourse that approaches ecological issues in their global, regional, and national dimensions and in their relationship to economic and social development. In many occasions, the Monarch took the opportunity to tackle this phenomenon, for example, during the national debate on the drafting of a Charter for the Environment in 2010, emphasizing the importance of incorporating the environment into government action and the need to drive the sustainable development process, making sure ecology plays a central role (Belhimer et al., 2016, p. 425).

Moreover, the speeches are based on a humanitarian ecology that makes humankind in the center of its interest. In this sense, His Majesty King Mohammed VI has based his vision on a new approach that revolves around consultation, continuous and serious search for practical solutions between all concerned parties, to ensure fairness, transparency, and the responsibility of each party according to their position and influence in global environmental issues, which cannot be achieved without effective partnership between states, civil society, the private sector, and international and regional organizations (Inter-ministerial Delegation for Human Rights, 2016).

The focus on environmental discourse reflects Morocco's position by virtue of its geographical location, cultural heritage, and history. Additionally, its strategic choices in the fields of environment and development are viewed in Morocco's involvement in the path of confronting and resolving various global problems and issues, including in particular climate change, water crisis, desertification, and biodiversity. Thus, the royal discourses have contributed to the implementation of the international conventions' principles signed and ratified by Morocco in the field of environmental protection.

However, in his speeches, His Majesty King Mohammed VI has strengthened the role of Morocco in dealing with the topic of environment. In this regard, he claimed in his speech addressed to High-Level Segment of COP 22, "Our country is among the first to have played a part in raising global awareness about climate change. Our initial contribution goes back to when I attended the Rio Earth Summit in 1992" (Albawaba Alwataniya, 2016). The speeches also highlight Morocco's achievements in the proactivity of sustainable development and the enhancement of environmental protection, such as the Environment Charter, the Green Morocco Plan, the Green Investment Plan, the ban on GMOs, and the new law on plastic waste, among others (Albawaba Al Wataniya, 2015). Moreover, the royal speeches have given special attention to the African continent, with more emphasis on Africa's suffering from climate change, natural disasters, and diseases, "The consequences of climate change are affecting developing nations as much as—if not more than—developed countries, especially the least advanced African and Latin American States and small island States." (Albawaba Al Wataniya, 2015). The diagram below represents the royal discourses that tackle the issue of environmental protection in national and international events.

As described in the aforementioned Figure 7.2,[2] the topic of environmental protection has been discussed in many occasions by His Majesty King Mohammed VI following the global needs to preserve the earth. Accordingly, Morocco has enacted many laws in this regard as presented in the following section.

Legislative Measures to Protect the Right to a Healthy Environment

The Moroccan legislator has explicitly enshrined the necessity of preserving the environment in the Constitution of 2011, and stipulated several laws that govern the field of environment from different angles. In this vein, this section attempts

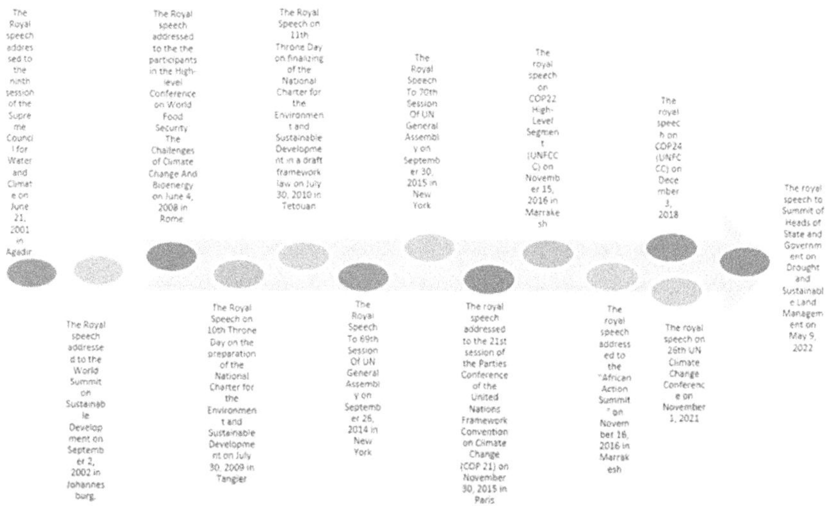

Figure 7.2 Royal speeches on environmental protection at national and international events. Source: Kingdom of Morocco: https://www.maroc.ma/en/royal-speeches.

to, first, discuss the main Articles related to environmental protection in the Moroccan Constitution, second, to introduce some of the main laws enacted to ensure this protection.

Constitutional Recognition of the Right to a Healthy Environment

In the last decades, the social and political movements in Morocco have led to the adoption of a new Constitution in 2011, aiming to address many national issues that have local, regional, and international dimensions. The constitutional legislator obligated the state, public institutions, and territorial collectivities to use all available and possible means to facilitate the benefit of equal treatment, healthcare, social protection, adequate housing, access to water, and the right to a healthy environment, among other provisions. The state is also working to achieve sustainable development that promotes social justice and equal opportunities, preserving national natural resources and the rights of current and future generations (Yahya, 2010).

Hence, one of the main topics tackled by the Constitution is the environment, which is considered as "an urgent matter on a regional and global level and a theme that the law has seized upon for various reasons, taking account of ecological imperatives but other factors too" (Belhimer et al., 2016, p. 425). Additionally, climate change in Morocco has experienced significant changes, mainly lower rainfall and higher temperatures in different regions, urging Morocco's public policies to take proactive actions, as represented in the constitutional amendment of 2011.

In its Articles, the Moroccan Constitution adopted a set of rules that explicitly or implicitly defend the environment, mainly Articles 19, 20, 31, 40, and Article 71 thereof. However, the reason for stipulating the 2011 Constitution on the right to a healthy environment is due to its interlink with the fundamental rights of citizens, such as the right to life, development, and human rights in general. Any violation of the environment is a violation of these rights (Yahya, 2010). Referring to Chapter 19 of the Moroccan Constitution, it states in its first paragraph:

> The man and the woman enjoy, in equality, the rights and freedoms of civil, political, economic, social, cultural and environmental character, enounced in this Title and in the other provisions of the Constitution, as well as in the international conventions and pacts duly ratified by Morocco and this, with respect for the provisions of the Constitution, of the constants [constantes] of the Kingdom and of its laws. (Morocco. Const. art 19, § II)

The Moroccan Constitution makes the rehabilitation of all segments of society and the constitutionalization of equality in rights and duties among its priorities (Lahrach, 2012). Thus, Article 20 maintains that "The right to life is the first right of any human being. The law protects this right." (Morocco. Const. art 20, § II). The right to life is considered one of the natural rights granted to every human being and imposes itself intuitively and does not require diligence. Therefore, all modern Constitutions stipulate the right to life along with the right to freedom and security. Accordingly, we find that the right to live in a healthy environment is closely linked to the right in life, since the existence of a healthy environment guarantees the right to life, and the right to life requires a healthy environment. The Constitution, in Article 31, stipulates that:

> The State, the public establishments and the territorial collectivities work for the mobilization of all the means available [disponibles] to facilitate the equal access of the citizens [feminine] and the citizens [masculine] to conditions that permit their enjoyment of the right:
>
> [. . .]
>
> to the access to water and to a healthy environment;
>
> [. . .]. (Morocco. Const. art 31, § II)

As for Article 40, the constitutional legislator elevates the culture of solidarity to a constitutional rule. According to the Article, solidarity is an expression of integration within a society or between a group of people and their surroundings, and the legislator requires all citizens to bear and support—"in solidarity and proportionally to their means—the expenses that the development of the country requires, and those resulting from calamities and natural catastrophes" (Morocco, Const. art. 40, § II), making solidarity among the various components of Moroccan society a vital field for social policies (Lahrach, 2012). However, Article 71 from the Moroccan Constitution declares that:

> [The following] are of the domain of the law, other than the matters that are expressly devolved on it by other Articles of the Constitution:
>
> [. . .]
>
> the rules relative to the management of the environment, to the protection of the natural resources and to lasting development;
>
> [. . .]. (Morocco. Const. art 71, § IV)

Consequently, the Parliament is considered the official body that is constitutionally competent to enact legislation and laws in general. The constitutional legislator then gives it the right to propose laws, discuss them and vote on them with rejection or acceptance, as well as promulgate them, among which we find rules relating to the environment, protection of natural resources, and sustainable development. The sixth section deals with the main laws enacted in the field of environmental protection.

Legislative Enactments to the Right to Healthy Environment

The Moroccan legislator has enacted a set of laws concerned with environmental protection, among which we find the following:

- Law N° 10-95 on Water in 1995
- Law N° 11-03 on the protection and improvement of the environment in 2003
- Law N° 12-03 on environmental impact studies in 2003
- Law N° 13-03 against air pollution in 2003
- Law N° 28-00 concerning waste management and disposal in 2006
- Law N° 57-09 on the creation of the Moroccan Agency for Solar Energy in 2010

- Law N° 22-07 on protected areas in 2010
- Law N° 22-10 on the use of degradable and biodegradable plastic bags and sachets in 2010
- Law N° 16-09 to turn the Centre for the Development of Renewable Energy into the National Agency for the Development of Renewable Energy in 2010
- Law No.° 13-09 on renewable energies in 2011
- Law N° 47-09 on energy efficiency in 2011
- The framework Law N° 99-12 on the National Charter on the Environment and Sustainable Development in 2014
- Law N° 58-15 amending and supplementing; Law N° 13-09 on renewable energies in 2015
- Law 81-12 on coasts in 2015 (Belhimer et al., 2016).

However, this section focuses on Law 11-03; Law 13-03; Law 77-15; Law 23-12; Law 99-12, and some insights from Forest Conservation and Exploitation Law, Money laundering Law, and Anti-Terrorism Law in covering the topic of environment. The environmental legislation in Morocco has been updated by adopting several new laws and amending existing legislation. The country has made progress in enhancing its environmental policy, with a focus on water management issues. Notably, three important environmental laws were approved in 2003, namely laws on the protection and conservation of the environment, combating air pollution, and environmental impact assessment.

Morocco's climate-change-related laws, policies, and plans recognize the importance of protecting the environment for the betterment of present and future generations. The country has also enacted laws concerning waste management and disposal, renewable energies, energy efficiency, and the creation of the Moroccan Agency for Solar Energy. The Moroccan government has prioritized environmental protection and sustainable development through the enactment of laws addressing water, forests, fisheries, and the management of natural resources.

Law on the Protection and Improvement of the Environment

The Moroccan legislator has enacted a set of laws concerned with environmental protection, among which we find Law N° 11-03 on the protection and improvement of the environment in 2003. This law introduces a set of articles aiming at protecting and enhancing the environment. Hence, in article 1, the legislator states the main objective of this law, which is to follow and decree the main national policy rules and principles in protecting and enhancing the environment. These principles include protecting the environment against all forms of pollution and degradation of whichever origin, improving the environment and living conditions of man, defining the basic orientations of the legislative, technical, and financial framework concerning the protection and management of the environment, and setting up a specific regime of responsibility that guarantees the restoration of damages to the environment and compensation for the victims (Law 11-03, 2003, art 1, chapter I, § I).

According to Article 2, the implementation of the provisions of this law is based on several key principles. These include the protection, improvement, and good management of the environment as integral parts of the integrated policy for economic, social, and cultural development. The protection and improvement of the environment are considered to be of public utility and constitute a collective responsibility that requires participation, information, and the determination of responsibilities. Additionally, the law emphasizes the necessity of balancing the requirements of national development with those of environmental protection when elaborating sectorial development plans and integrating the concept of sustainable development into these plans. It also highlights the importance of considering environmental protection and ecological balance when developing and implementing town and country development plans. The effective application of the principles of "the user has to pay" and "the polluter has to pay" is crucial in the execution and management of economic and social projects and the provision of services. Lastly, the law mandates respect for international covenants in the field of the environment when developing plans, programs, and environmental laws (Law 11-03, 2003, art 2, chapter I, § I).

Moreover, by examining the contents of the aforementioned law, we find that the legislator has mentioned two vital Articles; namely, Article 43 and Article 45, according to which they prohibit everything that may cause damage to health or the environment using the word "forbidden," and this indicates that we are facing peremptory rules that cannot be violated, because if they are exceeded, the perpetrator of acts contrary to the environmental field shall be subject to punishment.

It is also noted regarding the provisions of Law 11-03 that it came with a clear text on the objective responsibility, which states that the damage occurred even without proving the error in order to claim compensation for the damages suffered by people. This is indicated in Article 63, which states that

> Any individual or legal entity that stocks, transports or uses hydrocarbons or hazardous or dangerous substances, or any owner of a classified installation, as defined by the implementation texts of the present law, who have directly or indirectly caused bodily or material damage linked to the exercise of the above activities, are responsible even in the absence of proof of fault. (Law 11-03, 2003, art 63, chapter VI, § I)

Consequently, this law is also considered as the ground that draws the road map in the field of environmental protection and the preservation of the natural field in Morocco with regards to the implementation of other laws.

Law on the Prevention of Air Pollution

Air pollution, through the rise of Carbon dioxide in the earth's atmosphere, negatively affects air as well as green spaces such as plants and corps, and it is the main reason for global warming. Relatedly, Law 13-03 on the prevention of air pollution, states in its Article 2 that it aims to prevent and reduce emissions and air pollutants that could harm or affect "the health of man, the fauna, the soil, the climate, the cultural heritage and the environment in general" and this law is "applicable to every legal entity or individual that is subject to public or private law and that possesses, holds, uses or exploits real estate or mining, industrial, commercial or agricultural installations [. . .]." (Law 13-03, 2003, art 2, chapter II)

This law includes a set of articles through which the legislator aims to put in place the rules to protect the environment and prevent air pollution that would endanger human life. Article 13 of this law maintains that

> When air pollution is caused by an activity or an operation that constitutes a hazard for man and affects the vicinity, safety and the environment and if the hazards and damages were unknown and unforeseeable at the time of delivery of the permit or of submitting the statement of activity, the administration may issue to the person responsible for the source of pollution instructions to take additional measures or to introduce the necessary modifications to limit the emission of pollutants and to avoid the aforementioned hazards and damages. The administration orders that the necessary equipment and the available

> techniques be installed for measuring the concentration level of pollutants and their quantities as well as all the necessary means to maintain the authorized standards. (Law 13-03, 2003, art 13, chapter V)

However, the Article stresses that the administration has the right to order the cessation of the activity or exploitation of the source of pollution even if the responsible person uses all the necessary equipment to prevent air pollution (Law 13-03, 2003, art 13, chapter V). Moreover, the legislator, in Article 14, did not only deal with human health but the environment in general, and any threatening of the environment is a violation of this law (Law 13-03, 2003, art 14, chapter V).

The Conservation and Exploitation of Forests Law

Green space is considered one of the basic elements of living, given that it is a source of oxygen and the lung of any country. This fact is one of the main reasons why states establish and protect forests. In this regard, Morocco has adopted a policy of green belts in all regions and enacted a Law on Forest Conservation and Exploitation, which was issued in 1917. This law includes Articles that criminalizes acts that are contrary to this law and that would harm the environmental field in general and forests in particular (Almachichi, 2014). These Articles in the aforementioned law are 14; 23; 25; 31; 34; 35; 37 and 54 (Law of The Conservation and Exploitation of Forests, 1917, modified in 1960).

Law on the Management of Plastic Bags

Indeed, through Law 77-15, the Moroccan legislator has enacted a number of articles with a repressive nature, specifically Articles 9 to 13. These articles aim to ensure environmental considerations are taken into account by those working in the field of plastics, by invoking the environmental dimension in preventing the manufacture, import, export, marketing, or use of plastic bags. The Moroccan legislator, recognizing the environmental hazards posed by plastic bags, has enacted measures to prevent their proliferation and subsequent environmental impact.

The punitive aspect of this law includes substantial financial penalties, which can reach up to 500,000 MAD according to Article 10. This applies indiscriminately, whether the offender is a natural or legal person. As stated in Article 10, "Anyone who holds plastic bags referred to in paragraph 2 of Article 1 of this Law sell them, and put them on sale, or sell them or sell them distributed, for a fee or free, is punished by a fine from 10,000 to 500,000 MAD" (Law 77-15, 2015, art 10).

The law also criminalizes acts that threaten the environment in general and human life in particular, due to damages resulting from the spread of diseases caused by plastic bags. This is addressed through a series of articles contained in this law, namely Articles 9; 10; 11; 12 and 13. This underscores Morocco's commitment to environmental protection and sustainable development.

Waste Management and Disposal Law

The problem of waste is one of the difficult challenges that countries face, regardless of their position in the world, whether they are developed or developing countries. These challenges arise from the negative repercussions posed by diseases related to wastes, especially medical ones, which prompted countries to hold several international conferences to search for ways to minimize waste with less damage.

Waste has negative effects on the environment regardless of its type. For example, wastes can contribute to the rise of carbon, which affects the atmosphere and green spaces, in addition to the spread of dangerous diseases. Faced with this imminent danger, the Moroccan legislator drafted Law 23-12, as an amendment of Law No. 28-00, aiming at regulating the method of waste management and disposal. The Law includes a set of articles, part of which is devoted to injunctive provisions that criminalize a group of behaviors contrary to public order, and that entails punishment for the one who commits them (Komghar, 2008).

National Charter on the Environment and Sustainable Development Law

The framework Law 99-12 defines the basic objectives of the state's vision in the field of environmental protection and sustainable development. According to Article 1, this law aims to recognize environmental rights that should be protected and respected, focusing on the duties of the State, local authorities, public institutions, and companies concerning sustainable development. It seeks to strengthen the legal protection of resources and ecosystems by listing the types of actions or steps that the State proposes to take to fight against all forms of pollution. Additionally, the law aims to establish sustainable development as a core value shared by all segments of society and as a process followed by public policy development. It also aims to create a coherent and efficient system to implement the contemplated measures and lay down the groundwork for a system of environmental responsibility with a mechanism of financial compensation for damage to the environment. Lastly, the law seeks to establish environmental police to ensure respect for these rules (Eversheds Sutherland, 2013).

This framework law also stipulated a set of principles that must be adhered to when preparing and implementing policies, strategies, and programs by the state, territorial collectivities, public institutions, and state companies, as well as by other parties involved in the fields of environment and sustainable development. These principles are: principle of integration; principle of territoriality; principle of solidarity; principle of precaution; principle of prevention; principle of responsibility; and principle of participation (Law 99-12, 2014, art 2). This law also affirms the right of every citizen, male and female, to live in a healthy environment and gives them the right to participate in the process of making decisions that can affect the environment (Law 99-12, 2014, art 3).

Anti-Money Laundering in Protecting the Environment Law

The legislator did not hesitate to clearly criminalize the funds obtained from crimes committed against the environment based on the amendment to the Anti-Money Laundering Law. According to Article 574.1, it was considered that the proceeds from crimes against the environment constituted the material element for the crime of money laundering (Law 43-05, 2007, art 574.1). This provision comes in line with the global trend, which calls for developing internal legislation that can mitigate the negative impacts on the environment. This amendment has a crucial role in setting a preventive and repressive approach in dealing with environmental security since it includes the no impunity principle and allows recovery of funds of illegitimate origin (Adribila, 2015).

Fight Against Terrorism in Protecting the Environment Law

Law 03-03 on the Fight against Terrorism, released in 2015, considers environmental crimes as terrorist crimes if the motive for committing them is the destruction of the environmental field, and this is stated in Article 218.3, which considers this type of crimes as "ecological crimes." This Article maintains that bringing or placing a substance that endangers human or animal health or the environment in air, land, or water, including territorial waters, is considered a terrorist act, and these acts are punishable by imprisonment from ten to twenty years (Law 03-03, 2015, art 218.3).

Institutions and Projects Concerned with Environmental Protection

Morocco showed its commitment to preserve the environment by activating laws and conventions through the set of several projects and programs. This section sheds light on the main institutions and projects launched by Morocco to ensure healthy environment. If the legislative measures are an integral part of environment at all levels, then the institutional measures are the ones that ensure the embodiment of laws on the ground. This section intends to set the national institutions that work to preserve and improve the environment.

National Institutions: The National Council for the Environment

The National Council for the Environment, established thirty years ago, is considered the main platform concerned with embodying and activating coordination between all actors in the environmental field. This process takes into account the environmental concerns of all ministerial sectors, non-governmental organizations, and scientific institutions, and discusses in a cooperative way to improve the status of the environment.

The objective of the National Council for the Environment is to define the government's endeavors in the field of the environment, by activating, directing, and coordinating all actions related to environmental protection. Furthermore, it seeks to improve and manage the environmental affairs, encourage sustainable development by tracking studies and proposing draft laws related to the field of the environment, as well as educating and sensitizing the population through awareness campaigns (Ministère de La Transition Energétique et de Développement Durable, n.d.).

Economic, Social, and Environmental Council

The Economic, Social and Environmental Council emerged in the Moroccan Constitutional review of 1992, and then it was strengthened in the Constitution of 2011 within the institutional hierarchy of the state. Therefore, the council is considered as an independent constitutional institution, according to Article 11, due to the growing environmental dimension in dealing with the international and national political issues and it is framed in three Articles, namely, Article 151; Article 152 and Article 153. In this vein, Article 151 states the institutionalization of the council. Article 152 maintains that the council "may be consulted by the government, by the Chamber of Representatives and by the Chamber of Councillors on all the questions of economic, social or environmental character." And the council "gives its opinion on the general orientations of the national economy and of lasting development." (Morocco. Const. Art 152, § XI). Additionally, "It carries out advisory missions on major development guidelines, economic and social public policies as well as those concerning sustainable development and advanced regionalization" (Economic, Social and Environmental Council, n.d.).

Environmental Police

The Environmental Police was created according to Article 35 of Law 99.12, which stipulated the establishment of an environmental police with a mission to strengthen the authority of the concerned administrations in the field of prevention, control, and inspection (Law 99-12, art 35, § VI). Based on Article 35, Decree No. 2-14-782 was issued related to the organization and functioning of the environmental police. The first article specified the tasks of the environmental police in monitoring, inspection, investigation, research, and verbalization detection of environmental infringements, and editing records in this regard. Likewise, it intends to provide the necessary support to strengthen the authority of the administrations concerned with implementing environmental protection measures.

The main objective of establishing the Environmental Police is to identify the sectors and branches of activities that are of priority to subject them to environmental control, in view of the objectives of protecting the environment and preserving natural resources. Hence, monitoring areas and inspection activities are selected based on environmental stakes and risk identification. Besides, it also aims to prepare a long-term program to enhance human and technological capabilities and instruments to ensure these objectives; prepare, adapt, and simplify mechanisms and documents in the field of research; and investigate, inspect, and edit records regarding violations against the environment (Morocco, Decree No. 2-14-782, 2015, arts. 5–6).

Programs and Projects Enshrined in the Right to Healthy Environment

Within the framework of enshrining the right to live in a healthy environment, Morocco has undertaken a number of projects and implemented several programs aimed at preserving and protecting the environment. This section reviews the most important actions undertaken by Morocco in this context.

National Air Program

Morocco pays great attention to the problem of air pollution, which is mainly caused by fixed industries, means of transportation, among others, given its direct and serious impact on the health of citizens, especially children. To address this problem, the Ministry of Energy Transition and Sustainable Development prepared the National Air Program for Horizon 2030 in partnership with the concerned parties. The program was approved by the National Committee for Tracking and Monitoring Air Quality, and its axes focus on several key areas. These include strengthening and expanding the national air quality monitoring network, reducing gaseous emissions from the transport and industrial sectors, and strengthening the legal framework in the field of air pollution resulting from these sectors. Additionally, the program emphasizes the importance of strengthening communication and awareness in the field of air pollution.

Within the framework of activating this plan, the Ministry of Energy Transition and Sustainable Development took several measures, including strengthening the national air quality monitoring network by activating special agreements with the Kingdom's authorities, signing the agreement protocol and special agreements with the Mohammed VI Foundation for Environmental Protection, the Ministry of Interior Affairs and the General Directorate of Meteorology to enhance and develop this field (Ministère de La Transition Energétique et de Développement Durable, n.d.).

National Program of Waste Valorisation

The National Strategy for Sustainable Development has made the waste sector one of its axes by promoting the integrated management of waste. The management and development of the recycling sector is one of the steps to achieve a circular economy and the first step to resolve landfill. However, the management of household waste and similar waste began in 2007 through the National Program for Household and Similar Waste, by improving service and reducing environmental impacts while ensuring that waste is buried in accordance with international standards.

Since 2012, the environment sector has taken several measures in partnership with the concerned sectors to lay the main foundations for the development of waste valorization systems through the National Waste Valorization Program. This includes the establishment of appropriate legal frameworks and institutional arrangements that ensure good governance and the necessary financing in this sector.

The National Waste Valorization Program aims to lay down legal, institutional, technical, and financial foundations for the valorization and reuse of waste, by setting rules for an integrated and sustainable waste management, organizing waste recycling and valorization units. In addition to reducing waste of natural resources, minimizing the negative impacts of industrial activities, and improving the conditions of the national industry and creating investments and jobs (Ministère de La Transition Energétique et de Développement Durable, n.d.).

Zenata Ecological City

Zenata City is an innovative and sustainable eco-city designed to ensure well-being and good life for its residents. According to this vision, the city develops in harmony with nature, connects modern life with nature within the neighborhoods, and works to provide means of transportation with contemporary and modern specifications. Zenata City occupies a vast territory of 1,670 hectares located at the junction between Casablanca and Mohammedia. The city reflects natural diversity, as gardens and parks spread throughout the city and furnish the entirety of its sides. The city's park, which extends over an area of 5.5 km, is an invitation to enjoy the atmosphere of a picnic and to practice outdoor sports. The city is surrounded by sand dunes ranging in height from 8 to 12 meters, which form natural barriers in the face of waves that come from the ocean. Thanks to all these elements combined, this distinguished natural landscape gives the eco-city Zenata all the appropriate natural features that guarantee a better quality of life (Zenata Eco City, n.d.).

Conclusion

Morocco has witnessed an important economic leap by virtue of the development of vital socio-economic sectors, mainly agriculture, industry, fishing, urban development, and tourism, as well as infrastructure. Nevertheless, this leap had some negative repercussions on the environment. Consequently, in order to engage in the global responsibility campaign to protect the environment, Morocco has enacted a set of requirements aimed at achieving that goal. Therefore, Morocco has signed many international conventions that guarantee the right to live in a healthy environment. The Moroccan legislator has explicitly enshrined the necessity of preserving the environment in the constitution, which urges the state, public institutions, and territorial communities to put in place all available and possible means to facilitate the equal and fair right of a healthy environment for all members of society, in addition to the enactment of a set of laws that address some environmental problems and related issues. The efforts made by Morocco in this context are the greatest evidence of its real intention to search for common solutions, whether with regional or international actors, to reduce the risks surrounding the environment, taking into account the global development in different sectors. Morocco continues its efforts in the context of raising awareness about environmental protection, whether by launching programs or by including educational materials that develop a sense of citizenship among future generations and urge them to protect the environment. Finally, the chapter's main recommendation is to request the Moroccan legislator to collect all the laws related to the field of the environment in one law under the name of Environment Code in a way that facilitates research and reference. This collected law can also strengthen the state's efforts in this regard and make the laws clearer and more connected.

Notes

1 This figure contains some of the international conventions signed and ratified by Morocco in the domain of environmental protection. For more information about this topic, readers are invited to consult the following link: https://www.environnement.gov.ma/fr/lois-et-reglementations/conventions-et-accords.

2 For more information on the speeches we invite you to consult https://www.maroc.ma/en/royal-speeches and Interministerial Delegation for Human Rights *Alkhoutab wa Arasail Assamiya Almotaaliqa bi Albay'a* [Royal royal speeches and messages related to the environment], 2016.

References

Adribila, H. (2015). الركن المادي لجريمة غسل الأموال [The material element of the crime of money laundering]. *Majalat Achurta.*

Albawaba Al Wataniya. (2015, November 30). الخطاب الملكي الموجه إلى الدورة الحادية والعشرين لمؤتمر الأطراف. *[HM King Speech to COP 21].* Retrieved from Royal Speeches. Retrieved from: https://www.maroc.ma/ar

Albawaba Alwataniya. (2016, November 15). *Full speech of HM the King on COP22 high-level segment.* Retrieved from Royal Speeches: Retrieved from: https://www.maroc.ma/en/royal-speeches/full-speech-hm-king-cop22-high-level-segment

Almachichi, N. (2014). حماية البيئة البرية والجوية في القانون المغربي [Protection of the wild and air environment in Moroccan law]. *Majalat Alqanon Almaghribi*, 23.

Basel Convention on the Control of Transboundary Movements of Hazardous Wastes and their Disposal. (1989, March). 22: Retrieved from: https://www.basel.int/Portals/4/Basel%20Convention/docs/text/BaselConventionText-e.pdf.

Belhimer, A., Bras, J.-P., Dupret, B., & Mouaqit, M. (2016). When law addresses the environment: A matter of principle or a deep-seated concern? In M. L. Sabrie, E. Gibert-Brunet, & T. Mourier (Eds.), *The Mediterranean region under climate change: A scientific ipdate* (pp. 425–437). Marseille: IRD Edition Institut de Recherche pour le dévelopment.

Chalabi, K. (2015). المغاربي الفضاء في البيئي الامن تحقيق نحو: سليمة بيئة في الحق دسترة. [Constitutionalization of the right to a healthy environment: Towards achieving environmental security in the Maghreb space]. *Almajlis Adoustouri*, 63–87. Retrieved from http://www.webreview.dz/IMG/pdf/_-134.pdf.

Climate Change Secretariat (UNFCCC). (2005). *Proceedings of the seminar of governmental experts.* Bonn, Germany https://unfccc.int/resource/docs/publications/soge2005.pdf: CLIMATE CHANGE SECRETARIAT (UNFCCC).

Convention on Biological Diversity. (1992, May 22). Nairobi : Retrieved from: https://www.cbd.int/doc/legal/cbd-en.pdf

Dahir n ° 1-15-148 of 25 Safar 1437 (2015, December 7) bearing promulgation of Law No. 77-15 prohibiting manufacturing, import, export, marketing and use of plastic bags. (2015). Official Bulletin N° 6420 28 Safar 1437 (10 December 2015): Retrieved from:https://nicholasinstitute.duke.edu/sites/default/files/plastics-policies/1807_TR_Final_Morocco%202015%20plastic%20bag%20ban.pdf

Dahir N° 1-03-59 of 10 Rabii I 1424 (2003, 12 May) promulgating law N° 11-03 pertaining to the protection and improvement of the environment. (2003). P. 1900: Official Bulletin N° 5118. Retrieved from: https://www.climate-laws.org/geographies/morocco/laws/dahir-no-1-03-59-promulgating-law-no-11-03-on-the-protection-and-enhancement-of-the-environment

Dahir n° 1-03-61 of 10th Rabii (I) 1424 (2003, 12 May) promulgating law n°13-03 on air pollution. (2003). Official Bulletin 5118. Retrieved from: https://www.informea.org/en/law-no-13-03-prevention-air-pollution

Dahir n° 1-12-25 of 13 Ramadan, 1433 (2012, August 2) promulgating Law 23-12 amending Law No. 28-00 relating to waste management and disposal. (2012). Official Gazette No. 6078 dated Shawwal 11, 1433 (August 30, 2012), p. 4633.

Dahir N° 1-14-09, 4th Jumada Aloula, 1435 (2014, 6 March) bearing promulgation of Framework Law No. 99-12 on the National Charter for the Environment and Sustainable Development. (2014). Official Bulletin N° 6240 18 Jumada Aloula 1435 (March 20, 2014).

Dahir of 10 Octobre 1917 (20 Hijja) on the Conservation and Exploitation of Forests. (1917 modified in 1959 & 1960). Retrieved from: https://www.informea.org/sites/default/files/legislation/mor2830.pdf.

Eco City Zenata. (n.d.). *Eco City Zenata.* Retrieved from: https://zenataecocity.ma/

Economic, Social and Environmental Council. (n.d.). *ESEC Missions.* Retrieved from: https://www.cese.ma/en/

Eversheds Sutherland. (2013, June 24). *Morocco – The National Charter for Environment and Sustainable Development*. Retrieved from: https://www.eversheds-sutherland.com/global/en/what/articles/index.page?ArticleID=en/Africa_group/Morocco_Charter_for_environment_240613

Interministerial Delegation for Human Rights. (2016). **الخطب و الرسائل السامية المتعلقة بالبيئة** *[Royal speeches and messages related to the environment]*. Rabat: Manshourat Almandoubiya Alwizaria Almokalafa bi Hoqoq Alinsan.

Komghar, I. (2008). تدبير النفايات في المغرب بين الواقع والقانون [Waste management in Morocco between reality and law]. *Almajala Almaghribi Lil Idara Almahaliya wa Atanmiya*.

Kyoto Protocol to the United Nations Framework Convention on Climate Change. (1997, December 11). Retrieved from: https://unfccc.int/resource/docs/convkp/kpeng.pdf.

Lahrach, K. (2012). وتحليل شرح المغربية الدستور الجديد للمملكة - *[The new Constitution of the Kingdom of Morocco - explanation and analysis]*. Rabat: Silsilat Alamal Atachri'i wa Alijtihadat Alqadaiya.

Law No. 03-03 on the Fight against Terrorism. (2003). Official Bulletin of the Kingdom of Morocco, No. 5114 (5 June 2003). Promulgated by Dahir No. 1-03-140 (28 May 2003), art. 218-3.

Law No. 43-05 on Combating Money Laundering. (2007). Official Bulletin of the Kingdom of Morocco, No. 5522 (3 May 2007). Promulgated by Dahir No. 1-07-79 (17 April 2007), art. 574-1.

Ministère de La Transition Energétique et de Développement Durable. (n.d.). **الأنشطة المتعلقة بتنفيذ اتفاقية بازل** *[Activities related to the implementation of the Basel convention]*. Retrieved from : http://www.environnement.gov.ma/ar/strategies-programmes-ar/prevention-risques-ar/risques-chroniques-ar?id=346

Ministère de La Transition Energétique et de Développement Durable. (n.d.). **المجلس الوطني للبيئة** *[The National council for the environment]*. Retrieved from : https://www.environnement.gov.ma/ar/cne-ar/presentation-du-cne-ar

Morocco—Decree No. 2-14-782 of 30 Rajab 1436 (19 May 2015) on the Organization and Operating Modalities of the Environmental Police, Bulletin Officiel No. 6366 of 16 Chaabane 1436 (4 June 2015), arts. 5–6. Retrieved from https://dmp.uae.ma/textes_juridiques/generaux/decret_2_14_782_19%20mai_2015_police.pdf.

United Nations Framework Convention on Climate Change. (1992, May 9). Retrieved from: https://unfccc.int/files/essential_background/background_publications_htmlpdf/application/pdf/conveng.pdf.

Yahya, A. (2010). **التأطير الدستوري للبيئة في القانون المغربي والقانون المقارن** [Constitutional framing of the environment in Moroccan and comparative law]. *Almajala Almaghribiya Lilidara Almahaliya*.

8

Forest Policy and Management in Colonial and Post-Colonial India

A Historical Study

Prashant Tripathi

The contemporary concerns have led to an attempt to reconstruct Indian history using insights derived from recent debates in human ecology. The advent of colonial rule in India marked a vital event in the ecological history of India. Before the coming of the Raj, there were changes seen at the ruling authority's level (different rulers, dynasties), but not many changes at the rural level. The time before British rule in India has been called the golden age of ecological balance, notwithstanding the caste and class divisions, and had great coherence and stability. Despite modest cesses for specialized forest products like medicinal herbs, etc., indigenous tribes generally had unfettered usage of forests and wasteland in their area. However, there were no regulations placed on the vast majority of forest products utilized by the locals for their own food, livestock, and other household needs since they relied on the woods and their immediate surroundings for all of these things. The worship of holy trees and hallowed groves, as well as many local traditions and customs related to their natural environs, were only a few of the cultural and theological processes that codified their dependency (Guha, 1983).

But the scenario changed after the state control transferred to the hands of the British Raj. A profound transformation in the patterns of resource usage was taking place at home when the Europeans arrived in India. For our tale, three aspects of that transformation are very crucial. First, it placed less emphasis on gathering resources and producing food for survival and more on gathering, producing, transporting, and transforming resources for use as commodities. While fewer people were involved in food cultivation and subsistence hunting, more individuals were involved in manufacturing, transportation, or utilizing resources as commodities. Second, the importance of long-standing neighbourly cooperation, a trait of those who participated in subsistence food production, decreased with time. The result of this was the dissolution of strong local communities. Human society is fragmented, with people increasingly acting independently. Being that diverse, human groups were loosely organized with wide specializations and significant specialization crossovers, this was quite simple in Europe.

Lastly, and probably most significantly, the massive changes in resource use's "hardware" were matched by changes in its "software" that were just as significant. The ability of individuals to control access to resources was in high demand in the now-atomized society because of their potential for a limitless increase in resource usage. As manufacturing and business became the main economic activity, markets were used to organize resource availability. Thus, the devotion that food gatherers and Christian food producers reserved for spirits residing in trees was transferred to the institution of the market by the new set of beliefs that emerged. Success and prestige were now unambiguously expressed in terms of money, the market's unit of account (Guha & Ggadil, 1992).

It is impossible to grasp the whole scope of India and Britain's ecological interaction without first understanding these three features of the industrial method of resource utilization. For the majority of the Indian people whom the British were to dominate, the elevation of commercial needs above subsistence needs, the delegitimization of society, and the abandonment of limits on resource exploitation were all foreign experiences. Many different cultures and lifestyles collided here. Given the close link with Western colonialism and the destruction of nature that ongoing study has demonstrated, the environmental history of Colonial India is especially interesting. Western capitalism, whose dynamic growth caused damage to other organisms first by commerce and then by colonization, has significantly, changed the global environment. Such actions not only practically altered the environmental and sociological features of the ecosystems they invaded, but they also ensured that the modifications that resulted would largely benefit Europe. As mentioned in Webb (1964), the most obvious result of colonialism, one whose ramifications are being felt today, was its global monopoly on natural wealth.

A decline in the Victorian ethos of the businessman, which had allowed private parties to extract resources for significant benefit without concern for future generations, and a concomitant shift toward governmental involvement were necessary for the growth of environmental practice in India. An urge to reform the new industrial towns and enforce new standards of cleanliness, policing, education, and labor was made in response to the "utilitarian" theory of English political thinker Jeremy Bentham, who favored the general social good above unrestrained individualism. These changes represented a level of official interference unheard of in liberal Victorian culture. In response to this shift from Adam Smith's individualism to Benthamite collectivism, new environmental policies were implemented in British India, other British territories, and the United States (Perkin, 1989).

The Raj and the Forest Degradation: Early Phase

Around the year 1860, Britain had overtaken the United States as the leading deforestation nation. It was cutting down vast swaths of its own forests, in addition to those in Ireland, South Africa, and the northeastern United States, to use the wood in the construction of ships, the production of iron, and the cultivation of crops. The decimation of wood has occasionally been exploited as a political symbol by the British. As a result, the East India Company (EIC) destroyed the Ratnagiri teak forests that the great Maratha admiral Kanhoji Angre cultivated and flourished after defeating the Marathas in the early nineteenth century (Campbell, 1883).

Because of an absence of political will, prior conservation efforts in India had been fruitless. Even before the Napoleonic Wars began, in 1805, the British government asked the British East India Company, which already dominated huge swaths of the coastal areas, to look into the viability of harvesting Malabar teak in Madras to fulfill the demands of British shipbuilding. Even though the EIC was chartered as a business trading firm in 1600, it operated in India as if it were a government agency and held a near-monopoly on commerce within the regions it controlled. Together with Indian government officials, it acted as the British parliament in the country. In 1806, the company put Captain Watson, a former member of the Indian police force, in charge of the country's woods. Watson proposed a double strategy, one of which was to impose a tax on teak to reduce demand, and the other to limit its harvest by private merchants. This combination of techniques would prevent excessive harvesting and provide a continuous supply of teak (Ledzion, 1901).

A constant flow of strong wood for the Royal Navy was essential since "the stability of the kingdom hinged on its wooden defenses," and England's oak woodlands were rapidly disappearing (Stebbing, 1922). During a time of intense competition between colonial empires, India provided England with Indian teak, the most robust of shipbuilding timbers, which helped them win the war against Napoleon and later expand their navy. In addition to using teak that was brought into England, battleships were also manufactured in dockyards in Surat and around the Malabar Coast. Many expeditions were dispatched to the teak forests in an effort to locate the most promising sources of the bank (Edye, 1835). The British commercial fleet's tonnage increased from 1,278,000 in 1778 to 4,937,000 in 1860, demonstrating the rising demand for goods. Wood for the project was sourced largely from Britain's recently colonized territories. It wasn't until the 1880s that the British admiralty started pestering the Indian forest department for a steady supply of Madras and Burmese teak (Leathart, 1982). Deforestation was aided by the revenue-focused nature of British land management. Forests were seen as an obstacle to farming and thus a block to the development of the Empire, and their clearance increased the class of property charged for income. There was a general push in the direction of expanding agricultural production, and "the slogan of the period was to demolish the woods toward this purpose," thus the agricultural policy of the day reflected this priority (Ribbentrop, 1900).

The government's takeover of environmental assets, which had previously served as a shield for peasant families, was undoubtedly facilitated by forest conservation. However, colonial policies had additional, often unanticipated, ecological impacts. Irrigation projects in UP and Bihar caused certain regions to become waterlogged and resulted in soil salinization. Whitcombe argued that this placed an extraordinary burden on the Doab economic system. In Bihar, it was believed that the failure of locally owned public irrigation systems contributed to rural poverty and the emergence of Kisan Sabhas in the 1930s (Sengupta, 1980).

Railways and Deforestation

Environmental deterioration during British control in India has been linked to colonial rule in a number of studies. The British introduced railways to India in 1853 as an instrument of imperialism, hastening the rate of deforestation that had begun earlier. As a result of its widespread use, wood was in high demand for firewood, flooring, and even carriages. This resulted in random tree felling which occasionally left the landscape desolate. The foremost industrialized nation, Great Britain, lacked a "forest sensibility" and paid so little attention to forestry that she had to import 95 percent of her timber needs. As Britain's own wood supply ran out by 1860, she began destroying the woods of Ireland, South Africa, the North-eastern United States, Canada, and India to use in ship construction, iron smelting, farming, and occasionally as a gesture of political success (Troup, 1940). Evidence of a causal relationship between the British Raj and environmental deterioration in India has been uncovered by a number of studies. Even though deforestation had begun before 1853, when railways were introduced as a weapon of the British imperialism in India, the rate of destruction increased dramatically. They encouraged increased consumption thanks to convenient communication, leading to a surge in demand for wood for uses such as wood fuel, sleepers, and carriage construction. As a result, trees were cut down carelessly, leaving certain areas desolate.

With the advent of the railroad era beginning in about 1853, deforestation accelerated dramatically. A vast number of trees were cut down and left to decay on the ground while large swaths of woodland were cleared to fulfill the demand for railway sleepers without any oversight (Stebbing, 1922). For example, the forests of the sub-Himalayan of Garhwal and Kumaun region were devastated by the British; thousands of trees were felled down and those trees were not removed because it was impossible to replace them (Pearson, 1870). Railway firms also used wood from the forests to fuel their operations before the Raniganj coal mines were fully operational. Significant destruction of forest in the Doab was induced by the North Western Province's railways high wood fuel demands throughout the 1880s (Whitecombe, 1971). Cleghorn's book, *The Forests and Gardens of South India*, is one of the most evocative accounts of the natural landscape alteration caused by the railways. Once covered in forest, the grasping desire of the railways has now to a significant degree laid bare the slopes of Melghat and North Arcot. Wide swaths of ground have been cut away from the forest on all sides of the tracks, leaving no cover for any of the local wildlife.

To put it another way, marvelous changes on the face of the nation as respects tree vegetation were brought about by the development of the railway. More than 250,000 sleepers, or 35,000 trees, were needed each year in the Madras Presidency. Contractors increasingly relied on over-exploiting workers in sequence to supply this demand. The rainforest was cleared further away from the railway lines, and less desirable species were used up quickly, so they had to turn to others. More than fifty species were tested, but only a handful was ultimately deemed appropriate for use as railroad sleepers. Sleepers with a projected lifespan of five to six years only lived around a third of that. It was discovered that 458 (or 92 percent) of the 487 sleepers in a single shipment were made using illegal materials (Cleghorn, 1860). There were only three Indian woods that were suitable for use as railway sleepers in their untreated form; these were teak, sal, and deodar. The problem had reached critical proportions. Sal and teak were widely exploited in the early years because of their proximity to railway connections in peninsular India; consequently, expeditions were mounted to the north-western Himalayas in quest of deodar woods. In the years after the establishment of the forest department, the deodar of the Sutlej and Yamuna valleys was quickly depleted. Between 1869 and 1885, the Yamuna woods alone provided nearly 6,500,000 deodar sleepers (Hearle, 1888). Rapid railway construction in India, from 1,349 kilometers of rail line in 1860 to 51,658 kilometers of line in 1910, and the damage it wrought, made it abundantly clear that India's forest resources were finite.

The needs of the railroad industry were the first and by far the most significant factor in the reduction of India's forest cover. Since the 1857 insurrection 'had now become the topic of forest conservation, a significant administrative matter,' the Governor-General had urged in 1862 for the creation of a department to ensure the steady supply of the massive needs of the various railway companies for sleepers (Trevor and Smythies, 1923).

Establishment of the Forest Department: An Analysis of the British Forest Acts

Lord Dalhousie served as governor-general from 1848 to 1856, and his accomplishments during that time earned him the nickname 'the great Proconsul' among historians (Young, 1959). On August 3, 1855, Lord Dalhousie, India's governor general, reversed the country's earlier laissez-faire policy by creating the India Forest Department and annexed enormous regions of the lightly inhabited territory. Foresters, fireguards, rangers, and administrators now watch over these protected regions. In the decades that followed, forestry in India expanded into a worldwide field, ruled by experts from all over the world (Stebbing, 1922). Due to a lack of political will, earlier conservation efforts in India have failed. While the British East India Company had owned wide swaths of the coast, the British government asked them in 1805 to see whether they could harvest Malabar teak in Madras to fulfill the demands of British shipbuilding during the Napoleonic conflict.

In 1864, the Germans, then the European leaders in forest management, helped establish the imperial forest department. Prior to his posting in India, Dietrich Brandis, the inspector-general of forests, was the first to study botany at Bonn University. Stopping the massive deforestation of recent decades has necessitated, first and foremost, the creation of legislative instruments to establish and maintain governmental sovereignty over forests. In this double meaning, the railroads were a turning point for forest management in India, prompting the establishment of a dedicated agency (Guha & Gadgil, 1992). The Indian Forest Act of 1865 was the first attempt to establish governmental dominance. Thirteen years later, a far broader law took its place. However, the later act (which eventually became a template for forest regulation in other British-ruled territories) was passed only after a protracted and heated dispute within the British administration. This discussion, described in some detail below, has a particular significance for us today: the protagonists of the previous debate put out ideas remarkably similar to those given by participants in the modern discussion concerning the environment in India.

The 1865 act was hastily crafted in order to speed up the process of acquiring forest land set aside for railway supply purposes. It aimed only at staking the state's claim to the woods it urgently needed, with the caveat that no pre-existing rights be diminished in the process. The hunt for a more comprehensive and harsher piece of law began very shortly. In 1874, forest officials met for a meeting at which they discussed the flaws in the 1865 statute and the specifics of a new law. Following the 1875 meeting, Brandis drafted a paper on forest legislation based on the discussions there. The Indian Forest Act of 1878 was the result of more work on the earlier joint declaration by Brandis and a senior civil servant, B. H. Baden-Powell. When it comes to the definition, administration, commutation, and extinction of customary rights by the state, Brandis argues that Act VIII of 1865 is deficient in many areas. There was a lot of back-and-forth in the dispute about how to implement this division of rights, and three main camps eventually formed. The first group, who we refer to as annexationists, wanted complete governmental authority over all forests. In the second, more pragmatic, view, supporters of state administration of environmentally fragile and strategically important woods say that other areas should be left to communal forms of management. Third, the populist viewpoint is a reflection of the first. It argued that local communities, such as tribes and peasants, should have unfettered control over forests without interference from the state.

The sociology, history, politics, and ecology of forest resource use aligned with these three viewpoints on state control. They should be revisited in their entirety, since the problems they addressed—and argued so passionately a century ago are still with us today (Guha & Gadgil, 1992). The 1878 legislation was based on Brandis's 1875 memorandum and Baden-Powell's 1876 Forest Conference report. Drawing on Baden-Powell's distinction between "rights" and "privileges," the Act was a comprehensive piece of law that sought to erase centuries of customary use by rural inhabitants across India. It established three forest types.. "Reserved woods" were confined, valuable regions near settlements that could be used sustainably. In protected forests, a legal separation of rights was sought to ensure ultimate governmental control by extinguishing, transferring, or allowing restricted exercise of private rights. In 'protected woods' (also state-controlled), rights were documented but not established. Nevertheless, regulation was strictly enforced by the detailing of specific measures for the restriction and reservation of certain tree species as and when they became commercially useful, and for shutting the forest to pasture and fuel wood gathering whenever necessary. Due to rising commercial demand and the government's concern for their security, protected lands have been transformed into reserved forests.

As a result, by 1900, there were 56,000 square miles of reserved forests and 20,000 square miles of protected woods, compared to 14,000 square miles in 1878 (the year the legislation was established) and 81,400 square miles and 3,300 square miles, respectively, ten years after (Stebbing, 1922). The legislation also allowed for the establishment of village forests, a third type of woods; however, the government chose not to use this provision for most of the subcontinent. Lastly, by strictly controlling the drawing out and transportation of forest products and outlining a comprehensive range of punishments for violations of the acts, the new law significantly increased the administrative fines applicable to forest management.

Government dominance proponents knew full well that commercial wood extraction required stringent control (and ideally eradication) of traditionally held rights. Each 'rights holder' family was allotted a certain quantity of timber and fuel under the 1878 statute, and it was illegal to sell or barter any products originating from the forests. As a result, people were excluded from forest management on two fronts: physically, since they were forbidden access to woods and grazing, and socially, because they were given a weak and unyielding claim to the products of the forests. The significant forest policy declaration from 1894 was founded on the notion of governmental dominance. The plan was inspired by a damning government condemnation of the commercial orientation of forest management and implemented in response to "severe dissatisfaction among the farming classes" due to stringent regulations (Voelcker, 1893).

The Indian Forest Act of 1927, passed under the British colonial authority, aimed to regulate the harvesting of lumber from forests rather than to promote forest preservation. The British Empire established laws to further consolidate its colonial dominance in the same way it did with other forms of legislation. This historic statute, India's primary forest law, had little to do with conservation and instead continued the country's forest usage policy from 1878. To declare woods government property and utilize their timber was an attempt to subvert traditional rights and forest management systems.

A Purely Commercial Purpose

The Forestry Act of 1878 laid the groundwork for "scientific" forest management, which allowed for the clearing and harvesting of densely forested areas for industrial wood output while still enabling the government to exert tight control over forest use. The government's 99,000 square miles of protected woods were managed in accordance with imperialist objectives, beginning with the age of railway construction and continuing through both world wars. However, it was a continual effort to locate buyers for the many species found in India's tropical forests, of which only a small percentage could be sold. Many forest regions are inaccessible, and industrial growth has been slow, both of which reduce the potential for profitable exploitation of forest products. (Smythies, 1925)

As can be seen in Table 8.1, the division routinely posted a sizable excess of funds. Fuel wood, furniture, construction lumber, etc., were all in high demand in metropolitan areas, and the better connectivity created by the railway network made this feasible. In this way, the Himalayan woods supplied various kinds of coniferous forest to the urban markets of Punjab and the United Provinces, as well as to the military residential areas and hill stations that were established during the British administration. Even though the state had taken full control of the railways, it remained a major buyer of timber. Because in many situations the objectives of the two departments are equal, one committee urged better cooperation (Guha & Gadgil, 1992).

Table 8.1 Revenue and Surplus of Forest Department, 1869–1925 (Stebbing, 1922)

Yearly Average for the Period	Revenue (Rs/Million)	Surplus (Rs/Million)	Percent of Column 3 to Column 2
1869–70 to 1873–74	5.6	1.7	30
1874–75 to 1878–79	6.7	2.1	31
1879–80 to 1883–84	8.8	3.2	36
1884–85 to 1888–89	11.7	4.2	36
1889–90 to 1893–94	15.9	7.3	46
1894–95 to 1898–99	17.7	7.9	45
1899–1900 to 1903–4	19.7	8.4	43
1904–05 to 1908–09	25.7	11.6	45
1909–10 to 1913–14	29.7	13.2	45
1914–15 to 1918–19	37.1	16.0	43
1919–20 to 1923–24	55.2	18.5	34
1924–1925	56.7	21.3	38

Aside from lumber, the twentieth century also saw the cultivation of so-called "Minor Forest Produce" (henceforth MFP). The MFP was discovered to have several applications in manufacturing. Some of the most precious MFPs, like resin and turpentine, tanning ingredients like Kath and Myrobalans, and essential oils, could only be sourced from India, leading to a dramatic increase in exports. In 1917–18, for instance, the value of the lacquer export trade was estimated at about £2.5 million (Smythies, 1925).

The vast forested territories controlled by India's kings at the time were also dragged into the sphere of British capitalist growth, although peripherally. When these woods were exploited, either by the British authorities or by the princes themselves, it fundamentally altered the power dynamic between the two authorities. Over time, indigenous governments were usually very much awake to the significance of their forest assets. The colonial state's interest was first drawn to the Himalayan state of Tehri Garhwal in 1865 by the sight of its spectacular deodar woods; the following year, it secured a contract on the region's valuable Deodar and Chir forests. From 1865 until 1925, the Imperial Forest Department privately maintained these woodlands after removing the current privileges of the user of the nearby community. Although the per-year earnings were over Rs 1.6 lakhs (1910–55), the monarch did not extend the leasing and began managing the woods personally, with assistance from the forest department and under laws structured after the 1878 statute, due to their immense significance. The colonial authorities also contracted in the deodar woods of the neighboring hill regions of modern-day Himachal Pradesh, since deodar was valued everyday more and more for sleepers. The majestic Sal woods of the Himalayan province of Sikkim were also the object of the British's envious gaze. As time went on, various states in Central India, including Rewa, rose to prominence as sources of MFPs like Lac (Smythies, 1925).

The World Wars emphasized firmly the strategic importance of India's forests, which was initially shown during the construction of the railway network. The erection of bridges, docks, residences, shelters, and battleships during the 1914–18 war required the supply of wood and bamboo. 50,000 tons of pasture grass were sent to support military operations in Egypt and Iraq, while 228,076 tons of lumber (excluding railroad sleepers) were provided by the armaments boards' newly established "timber section" in just less than one year (1917 to 1918). From 1914 to 1919, wood exports totaled around 1.7 million cubic feet per year, the majority of it teak. The woodlands of the Indian subcontinent took a particularly heavy hit during the Second World War. It was early in the year 1940 that Delhi established a timber department to coordinate the flow of wood products from the regions. For a long time, the Middle Eastern theater, and subsequently the Allied troops in Iraq and the Persian Gulf, relied only on lumber sourced from India. Woods had to be discovered to replace the few handfuls that had previously dominated the market, and this task became an "ultimate aim" due to the war (Guha & Gadgil, 1992).

The Collapse of the Artisanal Industry

By limiting the availability of traditional supplies of raw resources, British forest management led to the fall of many types of artisanal manufacturing, in addition to its obvious effect on the farming classes. Most prominent among these was bamboo, a material essential to several facets of rural existence. This plant of immense local benefit was at first treated as a weed by British foresters; early programs proposed its destruction from timber-producing regions, despite its widespread use in building works, basket weaving, household equipment, and musical instrument production, as well as food and feed for animals. When it was realized in the initial decades of this century that bamboo made excellent raw resources for paper production, there was a dramatic change on the part of foresters, who now actively promoted commercial use while retaining limits on village usage. A great number of weavers were now compelled to purchase their bamboo from either government-operated storage or the free market. In addition, restricted supply resulted in new types of social strife among the populace who relied on agriculture for their livelihood. Therefore, the Baigas, who had previously reinforced their slash-and-burn agriculture with bamboo weaving, were unable to maintain this secondary source of earnings after the Basors, an artisanal caste that specializes in basket work, affirmed their "workers union" authority to domination on bamboo provided by the forest department. This resulted in the Baigas losing this source of earnings from the Basors' actions (Prasad & Gadgil, 1981).

The production of charcoal-based iron, however, failed under colonial control, whereas bamboo acquired either clandestinely from the forest or legally from the market continues to play an essential part in modern village culture. Once again, we owe much to Verrier Elwin for his insightful research into the latter years of the business. Elwin's book about the Agaria, an iron-smelting tribe of the Central Provinces, includes a terrifying account of how a combination of high taxes on smithies works and a limited supply of charcoal resulted in a precipitous decrease in the number of active furnaces (from 510 to 136 between 1909 and 1938). Despite the fact that villagers were partial to the pliable iron produced by village smelters, the Agarias were soon out of the market as new technologies and more trade with Britain rendered their product obsolete. The Agarias were dedicated to their trade and did everything they could to fight back, including breaking forest rules prohibiting charcoal burning and seeking refuge in neighboring chiefdoms, where they were treated better. This decline, attributable to scarce fuel supplies and international competition, was confirmed by Dietrich Brandis, the first inspector general of forests, whose exhaustive examination of Madras Presidency uncovered the same trends (Elwin, 1942).

Both the depletion of local reserves of raw materials and the introduction of cheaper mass-produced imitations from elsewhere in the world spelled doom for other types of artisanal manufacturing. From the 1870s forward, the entire country of India saw a decrease in the Tassar silk business, which had relied on the harvesting of wild cocoons from the jungle. The new forest rules may also be to blame for this deterioration, in this case, because of the higher taxes imposed on weavers who harvest cocoons from the forest. The domestic sector couldn't contend with the newly-established center of production running from towns, but the Tassar business eventually witnessed a renaissance under state sponsorship (mostly in reaction to a booming export market). Village industries and crafters have also declined as a result of a lack of availability of forest-based raw goods (Guha & Gadgil, 1992).

Nomadic is one community about which we know relatively less, but who would have been negatively impacted by forest rules? Unfortunately for the Gaddis, who live in the northwest Himalayas (modern-day Himachal Pradesh), the formation of restricted forests prohibited them from using the land to graze their sheep. They tried taking their sheep to pasture in the grassland waste and restricted woods but discovered that the groups of established farmers who owned these regions were hesitant to let their sheep pasture there. As a matter of fact, the Gaddis were opposed not only by the villagers but also by the forest authorities, due to the erratic nature of the pastoral cycle. The growing demand for Gaddi meat and wool in the lowlands coincided with tighter regulations on grazing, which paradoxically encouraged Gaddis to raise their group sizes (Guha & Gadgil, 1992).

Forest Policy and Management in Post-Colonial India

It is widely agreed that the unpredictable history of forest policy in post-independence India, with its shifting priorities, revised strategies, and ongoing improvisation, is both fascinating and significant because it focuses on key concerns of contemporary societal transformation (see Table 8.2). These concerns are typically categorized under the headings of environmental, technical, and collaborative issues of local communities. However, they also encompass apprehensions and uncertainties about income and justice, structural and scientific approaches, grassroots basic local-level democracy, and bureaucratic understandings (Bandopadhyay, 2004).

After India gained independence from the British, the new government sought to examine the previous administration's forestry policies. The Indian government decided to revise its strategy because dramatic developments that had taken place over the interim in the physical, economic, and political domains called for a reorganization of the existing policy. They predicted that forestry would grow in significance in India's economy and that forestry policy would become a major factor in diplomatic ties between countries. In 1952, the government of India passed a policy that served to be the first official articulation of the country's forestry legislation since independence. While the new national plan was supposed to be reflective of modern realities, it was primarily based on the 1894 Forest Policy Act. The 1894 policy declaration was affirmed to be "the forest policy of India to this day" and to have "basic ideas . . .

Table 8.2 Chronology of the Forest Policy

Date	Policy/Acts	Properties
1952	National Forest Policy	Expanding the 1894 Forest Policy on the grounds of "primary national need."
1970	National Commission on High demand and production of timber	Agriculture for the commercial & industrial use.
1980	Forest Conservation Act	Limits the ability of the state govt. and other bodies to act independently without central government approval.
1988	National Forest Policy	Focus on conservation, sustainability people's participation.
2006	Forest Right Act	Deals with the rights of forest dwellers, tribes, and their needs.

still remain valid" in a law enacted in 1952 (The National Forest Policy of India, 1952). In the pretext of the country's interests, the 1952 policy just established monopoly rights at the expense of the people living in forests. Furthermore, the specified goals lacked a defined hierarchy, which left the interconnected concept of forest usage unmodified. Forest-based businesses expanded quickly in India after it attained independence. Large-scale planting of rapidly expanding, high-yielding species of trees became the norm starting in 1959. Within the first three decades of independence, the paper industry saw unheard-of expansion. Within a decade, from 1966 to 1977, the production of industrial wood surged by more than 50 percent. The unprofitable conservation-focused tactics were abandoned (Guha, 1983).

As productive forestry for the production of commercial and industrial wood, the National Commission on Agriculture (NCA) described the conservation approach. The organizational structure of the forest was significantly altered by the forest authority since the conventional approach of maintained yield is incompatible with the companies' expanding needs. Slow-growing trees are being removed to make room for species with fast growth (National Commission on Agriculture, 1976). A refocused forest policy's primary objective for the NCA was to substantially increase the rate and scope of Indian forestry industrialization. The largest category of woods, which includes the best forests in the nation, would be industrial forests.

The Fifth Five-Year Plan era saw the immediate implementation of NCA's recommendations for forest management. Operations for clearing trees were started, and industrial planting projects were advanced significantly (Haeuber, 1993). In the resolution adopted at the XXII Central Board of Forestry (CBF) meeting held in December 1987, the first policy level decision for public participation in forest conservation & administration was made. The Prime Minister emphasized the need for efficient public engagement in forest management and conservation in his chairman's remarks. This was also noted in Resolution No. 25, which is as follows: This assembly decides that by 31.3.90, each village community will have an ecological protection and restoration and forest revival plan. The complete involvement of local Panchayats or other similar organizations would be utilized in the development and execution of this strategy. The 1992 Rio Forest Guidelines and all subsequent sustainable strategies were covered under the newly enacted 1988 'Forest Policy.' The primary aims of this policy were restoring ecological balance, conserving soil and water, and maintaining natural stability, spreading awareness among the people about the protection of trees and natural resources, expanding the forest cover 33 percent on land and 66 percent in hilly ranges, economic aspect also must be promoted along with these principles (Mukherji, 2003).

In India, collaborations in the conservation of forests, including both state forest agencies and local people are known officially as Joint Forest Management, or JFM. The Indian National Forest Policy of 1988 is a full description of the goals and aims of the Joint Forest Management. Following their first achievements in West Bengal and Haryana, JFM programs were given national prominence in 1988 law and a push in 1998 guidelines. The collaborative management of more than 1400,000 km2 of forested land was being carried out by over 63,000 FPCs in twenty-seven states of the Indian Union as of the year 2000. A total of 2460000 km2 was administered in 2010 by more than 112,896 committees, and the JFM program benefited almost 14,500,000 households.[1]

Forest Rights Act, 2006: Glimpse of Hope

The Forest Rights Act of 2006 is a call to account for those underprivileged people whose natural habitats were abruptly taken from them. Without giving it a second thought, they knock off the justice's splendor, clip the wings of the kite that declared the genuine homes of those scheduled tribes to be "reserved forests," and sever the mutually beneficial tie that has existed since the dawn of civilization. It re-establishes the intricate link between tribal and forest life. The tribe's rights are recognized, as are those of the forest inhabitants whose very existence depends on the forest; in exchange, these people defend the forest as their mother. The development of a local culture of tribes, which once again receives nourishment from this law, is the secret value underlying this reciprocal interaction. By giving the tribe, allow resources again and granting them the freedom to live their lives and earn a living according to their needs, it restores the tribe's rights while maintaining the biodiversity and habitats in accordance with their covert dedication to Flora and fauna. After the Forest Rights Act of 2006 was put into effect, several states still exhibit a desire to fully administer and advertise it since it boosts income.

The government has made a sincere attempt to be present and monitor the situation until a law is fully implemented. Otherwise, true progress cannot be carried out, no matter the cost, in its positive spirit. The opinions and attitudes of the forest agencies regarding the tribe need to change. To honor and utilize tribal knowledge, the monopoly of forest departments has to be broken and a cooperative relationship should be developed (Pal & Souran, 2022).

A historic set of laws known as the Scheduled Tribes and Other Traditional Forest Dwellers (Recognition of Forest Rights) Act was approved by the Parliament of India after a contentious formulation involving several parties with competing interests. The Campaign for Survival and Dignity (CSD), a union and governmental framework of tribal and forest dwellers' associations, served as the primary catalyst for the proclamation by drawing focus to the condition of Tribal (46 percent of the tribal communities resided in extreme poverty, compared to 27 percent nationally, and 200,000 tribal families living across 5,000 rural communities and small villages did not have access to the territory they farmed for nourishment).

The legislation promised a more stable foundation for handling and proper administration of natural resources while addressing the "past injustices" by reinstating the rights of land-use (at the personal and group level) amongst Tribal people and other indigenous forest-dwelling people. This legislation Forest Rights Act (FRA) gave a ray of hope to the people who were relying on forests. On January 1, 2008, the Act went into effect, with the intention of protecting the lives and means of subsistence for 100 million of the poor people. For the gathering and selling of minor forest products (MFP), the Forest Rights Act of 2006 creates a legal structure (MFP). These consist of the Tendu leaf used in Bidi businesses and the very valuable bamboo that was originally under the control of the forest department. For the indigenous residents of these settlements, claiming control of these riches meant achieving financial freedom. Only 46,156 out of the 2.9 million cases that were resolved under the FRA involved community rights, based on a 2010 media statement from the ministry of tribal affairs, and the most of these did not entail claim over MFP (CSE, 2018)

The Draft Indian Forest (Amendment) Bill 2019: Arming Forest Bureaucracy

The Forest Rights Act (FRA) of 2006 recognized and secured the rights of forest-dwelling Scheduled Tribes and other traditional forest dwellers in India, empowering them to access and use forest resources sustainably. However, the proposed Indian Forest (Amendment) Bill of 2019 seeks to restore state control over forests at the expense of such rights. The proposed law introduces a cooperative forest management committee to oversee "village woods" and other forest resources, but the Forest Rights Act supersedes all other laws in acknowledging and granting rights over forest land and resources to villages that live in forests, including the ability to conserve and control forests by their Gram Sabha, rendering the JFMC ineffective or unnecessary. The proposed bill also permits forest bureaucracy to use firearms, enter any building, and conduct an investigation on the basis of simple suspicions. Additionally, the federal government would have the authority to veto state laws and get involved in forest management. Experts have raised concerns that the proposed measure strengthens the authority of the forest administration rather than giving forest inhabitants the advantages the FRA promised. The Forest Rights Act is a vital law that acknowledges the rights of India's Adivasi people and recognizes their right to protect, regenerate, or conserve or manage any community forest resource, which they have been traditionally protecting and conserving for sustainable use.

Conclusion

The aforementioned narrative clearly shows that post-British forest policy is quite similar to British forest policy. Conflicting priorities have driven the creation and implementation of governmental policy. Prior to 1947, woods were a critical raw resource essential for colonial purposes like railroad construction and the world wars, but in post-independence, forest management has been governed by industrial and economic objectives. The successful execution of legislation, though, has only been accomplished in both instances at the price of the forest communities and their vital support systems. The disadvantaged groups have often reacted by using a variety of demonstration tactics, from somewhat unplanned eruptions and unrests to more structured protests.

Numerous manifestations of the organic connections between British and post-British forest policy exist. Firstly, in the contradictions among forest policy's "philosophy" and "practicing." Although the populist vocabulary of many forest law texts openly acknowledges and articulates the concerns of rural people, the execution of legislation has generally focused on recognizing the productive role of the woods. Second, over the two-time frames under consideration, forest administration and law utilized pretty stable techniques to achieve the objectives of forest legislation. Instead of considerable deviations from pre-existing colonial traditions, adjustments during the latter time have been primarily in the form of adjustments and expansions.

In terms of survival, the tribal communities and the forests are equivalent, and both have the same duty to look out for one another. They have a mutually supportive connection together. They rely on the forest for both their religious and mystical notions in addition to their financial basic needs. They revere different species of flora and animals, known as Totems, as their gods and deities. Preference must be given to those for whom the blue strategy of growth is intended, but in the context of tribal, this style of speaking was not heard. Since the Indian people are the only ones responsible for the nation's growth, it must be required that plans for that advancement be made on their own.

The entrenched interests of the officials have berefted the forest scene of its diversity and life under the guise of 'nature preservation.' A government regulatory framework that can play a crucial part in protecting supplies and resources that can support life in the combined interests of people and the environment should be made and implemented with the modest objective of maintaining ecological balance and tribal wisdom. Additionally, it is borne in mind that illegal trade cannot replace it. In order to continue conserving the environment and tribal rights, indigenous knowledge should be kept consistent and empirical. When a nation upholds the rights of its citizens, those citizens uphold the integrity and advancement of the nation.

Note

1 https://unacademy.com/content/full-forms/jfm-full-form/

References

Bandopadhyay. (2004). Three issues from a CPR management: Village forestry in post-colonial South Asia. In B. B. Chaudhuri & A. Bandopadhyay (Eds.), *Tribes, forest and social formation in Indian history* (pp. 209–224). Manohar Pubns.

Campbell, J. (1883). Bombay gazetteer: *Kanara.* Government Central Press.

Cleghorn, H. (1860). *Forests and gardens of South India.* W.H. Allen.

CSE. (2018). *People's Forests: CSR in Bamboo can drive sustainable livelihoods and socio-economic development at the bottom of the pyramid.* Green Golden Bamboo. Retrieved from: https://www.cseindia.org/people-s-forests-8540

Edye, J. (1835). Description of the sea-ports of the coast of Malabar, of the facilities they afford for building vessels of different descriptions, and of the produce of the adjacent forests. *Journal of the Royal Asiatic Society, 11.*

Elwin, V. (1942). *The Agaria.* Oxford University Press.

Government of India. (1952). *The National forest policy of India.* Delhi.

Guha, R. (1983). Forestry in British and post-British India: A historical analysis. *Economic and Political Weekly, 18*(44), pp. 1882–1896. Retrieved from: https://www.jstor.org/stable/4372653

Guha, R., and Gadgil, M. (1992). *This fissured land: An ecological history of India.* Oxford Scholarship Online Publication. https://doi.org/10.1093/acprof:oso/9780198077442.001.0001

Haeuber, R. (1993). Indian forestry policy in two eras: Continuity or change? *Environmental History Review, 17*(1), 49–76.

Hearle, N. (1888). *Working plan of the Tehri Garhwal leased forests. Jaunsar Forest Division.* Government Press.

Leathart, S. (1982). Review of N.D.G. James, a History of English Forestry. *Times Literary Supplement.*

Ledzion, M.M. (1901). *Forest families.* London.

Mukherji, A. K. (2003). Forest policy reforms in India – Evolution of the joint forest management approach. *World Forestry Congress*, Canada.

National Commission on Agriculture. (1976). *Report of the National Commission on Agriculture.* GOI.

Pal, S., & Souran, B. (2022). Forest and First Nation in post-colonial India. *AGPE the Royal Gondwana Research Journal, 3*(3).

Pearson, G. F. (1870). Deodar Forests of Jaunsar Bawar. *Selections from the Records of the Government of the North-western Provinces, 2nd series, vol. II,* Allahabad.

Perkin, H. (1989). *The rise of professional society* (pp. 363–365). Routledge.

Prasad, S. N., & Gadgil, M. (1981). *Conservation of Bamboo Resources in Karnataka.* Karnataka State Council of Science and Technology. Bangalore.

Ribbentrop, B. (1900). *Forestry in British India.* Government Press, Calcutta.

Sengupta, N. (1980). The Indigenous Irrigation Organization of South Bihar. *Indian Economic and Social History Review, 17.*

Smythies, E. A. (1925). *India's Forest Wealth.* Humphrey Milford, London.

Stebbing, E.P. (1922). *The Forests of India.* John Lane, London.

Trevor, C.G. and Smythies, E.A. (1923). *Practical Forest Management. a handbook with special reference to the United Provinces of Agra and Oudh. Allahabad:* Government Press, Allahabad.

Troup, R. S. (1940). *Colonial Forest Administration.* Oxford Publications.

Voelcker, J.A. (1893). *Report on the Improvement of Indian Agriculture.* Government Press, Calcutta.

Webb, W. (1964). *The Great Frontier.* University of Texas Press, Austin.

Whitcombe, E. (1971). *Agrarian Conditions in Northern India, Volume I: The United Provinces under British Rule, 1850–1900.* University of California Press, Berkeley.

Young, G. M. (1959). *The English utilitarians and India.* Cambridge University Press.

9

Moroccan Cinema

Sotto Voce as a Case Study

Mohamed Mliless, Latifa Belfakir, and Fouad Souiba

For several decades, the world has been facing an environmental crisis. Despite scientists warning about the urgency of the situation since the 1970s, it is only in recent years that the public has become more attentive. Various collective movements and associations are mobilizing to raise awareness and find sustainable solutions. In parallel, cinema and television are gradually catching up, with initiatives such as Film Climate and Ecoprod, among others, committed to a more environmentally respectful society. The goal is to leverage this influence to integrate the sector into the Moroccan national approach of achieving carbon neutrality by 2050.

This study aims to evaluate how cinema can contribute to this collective effort. We examine the evolution of Moroccan cinema, from early representations of nature in cinema to the activism of some filmmakers, establishing a link between the maintenance of the environment as a socio-political issue and the proliferation of ecological films.

Cinema has long been interested in the environment, exploring the relationship between humans and nature, the impact of human activity on the environment, and the consequences of environmental degradation. From documentaries exposing environmental issues to fictional narratives imagining dystopian futures, cinema has the power to shape our understanding of the environment and our place within it. This study aims to examine the representation of the environment in Moroccan cinema, exploring how films, such as *Sotto Voce* by Moroccan producer Kamal Kamal, have portrayed environmental issues and how these representations have evolved over time.

One of the most comprehensive books on Moroccan cinema is Carter (2009) titled *What Moroccan Cinema? A Historical and a Critical Study, 1956–2006.*

In this book, Carter recounts changes of a fifty-year period in the Moroccan legislation, institutions, supplementary influences, persons involved in the sector, representative films, and cinema culture. The book examines cinema as an institution in depth, addressing exhibition and distribution, censorship, and film clubs and caravans. Furthermore, the book is based on an examination of representative films from each era. Her unique study will be useful to anyone interested in film and Middle Eastern studies.

Moroccan cinema has a rich and diverse history, with a wide range of genres and themes that consider its transnational dimension (Martin et al., 2020). From historical epics to contemporary dramas, Moroccan movies reflect the country's unique cultural heritage and complex social landscape (Iddins, 2017; Carter, 2008, 2009; Filmmaking Lifestyle, n.d.). Themes explored in Moroccan cinema include poverty, religion, extremism, love, sexuality, politics, faith, tradition, and the human spirit (Filmmaking Lifestyle, n.d.). For example, Nabil Ayouch's film *Ali Zaoua: Prince of the Streets* (2000) tells the story of homeless children in Casablanca who embark on a dangerous journey to fulfil their friend's dream of a proper burial. Others have represented taboo themes in Moroccan cinema through analyzing one of its most famous contemporary films (Laabidi, 2022); Abdelkader Laktae's film *Love in Casablanca*, a brave initiation into tackling taboo topics in Moroccan cinema (Tbato, 2004).

About the Film *Sotto Voce*

According to the online French dictionary www.linternaute.fr, "sotto-voce" is an Italian term that means "under the voice." In music, the term refers to singing or speaking in a low voice, as if whispering. The term can also be used figuratively to indicate that someone is speaking very softly.

The film *Sotto Voce* by Moroccan director Kamal Kamal was screened at the opening of the 10th Arab Film Festival in Sao Paulo, Brazil. Made in a personal narrative style, the feature film retraces the struggle of Moroccans and Algerians against the French army during the Algerian War (1954–62). It tells the story of Moussa, a Moroccan resistance fighter charged with guiding groups of refugees fleeing the war through the mountains in the region of Beni Boussaid, on the Moroccan-Algerian border near Oujda. This Moroccan-Emirati co-production, with a duration of 1 hour and 34 minutes, features talented actors and artists including Younès Megri, Ahmed Benaïssa, Amal Ayouch, Jihane Kamal, and the late Mohamed Bastaoui. *Sotto Voce* is Kamal Kamal's third feature film after *Taif Nizar* and *Symphonie Marocaine* and has received several awards, including the Ousmane Sembene Grand Prize at the last African Film Festival.

The film's strength also lies in the theme chosen by the director, which is close to his heart, as it was part of his own life and family memories. Kamal (as cited in Le Matin, 2014) said,

> It's a bit of the story of my mother who made this passage, because she was a revolutionary soldier during the Algerian War. As she is Algerian, she was sentenced to death in absentia. So, we had her transferred to Morocco. She told us this story since we were children. That's how I got the idea to make a film about this passage from Tlemcen to Oujda.

But in *Sotto Voce*, this harrowing journey is accompanied by fabulous opera arias, in order to relieve the painful moments of suffering. Mohamed Khouyi, an actor in the film said (as cited in Le Matin, 2014):

> When I saw the film today for the first time like you, I liked it in terms of direction, acting, technicians, especially when I think of the difficult conditions in which we filmed, with the cold and snow. Now it remains the judgment of the public and professionals. For me, I did my best to live up to this role. I am happy and very proud to have participated. The most important thing is that the film pleases the public when it is released in theaters to contribute to the development of Moroccan cinema.

Sotto Voce as a Film About War

War films set in cold and snowy environments can create a powerful impact on the audience by heightening the tension and drama of the story and conveying the physical and emotional challenges faced by the characters. For example, *Sotto Voce* takes place in a snowy location and portrays the harsh conditions faced by a group of fugitives trying to escape the French army from Algeria to Morocco. The film's strength lies in its theme, which is close to the director's heart, as it was part of his own life and family memories. The director, Kamal Kamal, stated that "It's a bit of the story of my mother who made this passage, because she was a revolutionary soldier during the Algerian War." The film illustrates this moving story beautifully through symphonic music marked in *Sotto Voce.*

Some other examples of war films set in cold and snowy environments include *Into the White* (Næss, 2012), directed by Petter Næss, which is set in Norway during WWII and follows a group of British and German soldiers who find themselves stranded in the wilderness after an aircraft battle. They must put aside the rules of war to survive the winter in the same cabin (IMDa, n.d: IMDb, n.d.). These films use the cold and snowy environments to create a sense of isolation and vulnerability, increasing the emotional impact of the stories.

Overall, war films set in cold and snowy environments can create a more immersive and emotionally engaging experience for the audience by heightening the tension and drama of the story and conveying the physical and emotional challenges faced by the characters.

Review of the Literature

The environment has been a central theme in cinema since the turn of this millennium. Films have utilized the environment to explore a wide range of issues, including pollution, climate change, and the relationship between humans and nature.

In the literature review, the study will attempt to account for the following concepts: Arab cinema boasts a rich history and has produced many notable films. This section will delve into the history and development of Arab cinema. Moreover, it will examine the patterns and trends in new Arab cinema, including the emergence of Moroccan cinema. Moroccan cinema, recognized for its unique style and themes, will be discussed in terms of its history and development.

In addition to exploring the history and development of Arab and Moroccan cinema, this review will also scrutinize various themes prevalent in cinema, including the environment. Films that explore the role of nature in cinema and the carbon footprint of the cinema industry will be discussed. The representation of nature in cinema, its significance, and the environmental impact of the cinema industry will also be examined.

This section, while focusing on the 'visual as language' approach to cinema analysis, will explore the other approaches and will investigate how the major themes in war films, that is, landscape, nature, and environment, are depicted. It will also provide an analysis of Sotto Voice, a film that intertwines both the environment and war.

Arab Cinema

Arab cinema is renowned for challenging stereotypes and making political statements. Many countries experienced a surge in film production upon gaining independence (Euronews, 2021). Since the late 1980s, Arab cinema has highlighted concerns associated with the stagnation of the pan-Arab concept and the rise of Islamic extremism (Khan, 2001; Leaman, 2001; Sadoul, 1996; Shafik, 1988; Schirmer Encyclopedia of Film, n.d.).

Cinema is increasingly striving for a national identity that acknowledges its diversity and seeks a new role in the fight for social and national emancipation. In Egypt, Islamic fundamentalism directly impacted intellectual and artistic life, leading to the production of several films addressing the issue, such as Merzak Allouache's *Bab El-Oued City* (Allouache, 1994) and Yamina Bachir's *Rachida* (Bachir-Chouikh, 2002). Algerian and Tunisian filmmakers have also addressed fundamentalism, depicting its practices and influence on youth culture.

Popular films like Usama Mohammad's *The Box of Life* (Oussama, 2002) and Randa Chahal Sabag's *Le cerf-volant* (Chahal Sabbagh, 2003) have reflected on the region's colonial and neo-colonial rule. New Arab cinema emphasizes social and cultural situations and characters that portray a rapidly changing culture striving to reclaim its national identity amidst internal and external influences. The Palestinian issue remains a popular topic, with a focus on viewing the situation from the perspective of its real victims. Filmmakers such as Michel Khleifi, Elia Suleiman, Hany Abu-Assad, and Yousri Nasrallah have explored the politics of personal experience. The role of Arab Christians is a narrative thread in many new Arab films, underscoring the diversity of Arab identity and culture.

Increasingly, Arab filmmakers are exploring homosexual and bisexual relationships in Arab society. Examples include Daoud Aoulad-Syad's 1998 Moroccan film *Adieu Forain*, which features a gay transgender dancer in the lead role, and Nabil Ayouch's 2002 film *Une minute de soleil en moins* ("A Minute of Sun Less") (Ayouch, 2003), which features a main character whose partner is a transvestite. Other films, such as Akram Zaatari's documentary short *Diary of a Male Whore* (Abu Wael, 2001), are even more explicit in their defiance of sexual restraint, but due to their experimental nature, are less likely to reach a wide audience.

The evolution of Arab cinema is a pattern that has flourished due to advancements in communication technologies, including the growth of Arab satellite film and television networks, a significant factor in the expansion of Arab cinema at the end of the twentieth century. Importantly, film festivals in the region, such as Cairo, Beirut, Marrakesh, Damascus, Carthage, and Dubai Film Festivals, as well as the Ismailiah International Documentary Film Festival in Egypt, are growing and increasingly informing a renaissance of pan-Arab national cultural interaction.

Moroccan Cinema

Moroccan cinema benefited greatly from colonial cinema in its early stages, which provided Moroccan filmmakers with the techniques and tools necessary for film production. Numerous international films were shot in Morocco before local filmmakers began producing their own. In colonial films, Morocco was depicted as exotic, perpetuating a distorted view of the Arab continent as barbaric and primitive. The history of Moroccan cinema can be divided into three periods, starting with the year of independence: the first is from 1956 to the 1970s, during which filmmakers were employed as government workers. At this time, the Moroccan government recognized the power of film in fostering nationalism among Moroccans. The second phase spans from 1971 to 1985, when Moroccan cinema started to be influenced by the audience's demand for entertaining films.

Moroccan cinema is a relatively young industry, with its first feature film, *The Big Trip*, released in 1948. It quickly became a significant force in the Arab world, producing critically acclaimed films. Moroccan cinema is often characterized by its focus on social issues such as poverty, inequality, and corruption, as well as themes of identity and belonging. Notable Moroccan films include *The Battle of Algiers* (Pontecorvo, 1966), *Z* (Costa-Gavras, 1969), *The Silences of the Palace* (Tlatli, 1994), and *The Secret of the Grain* (Kechiche, 2007). These films have won numerous awards at international film festivals. In recent years, Moroccan cinema has seen a surge in popularity, both domestically and internationally. This is partly due to the increasing number of Moroccan films being produced, as well as the growing recognition of Moroccan cinema at international film festivals. Moroccan cinema is a vibrant and diverse industry that is continually evolving.

Movie Theme

Indeed, themes are the inferred stance taken on the subject of a story, and everything that happens within a story should reference back to a theme.

Anderson (2022) suggests that adopting a more philosophical and analytical approach to movie watching can help us understand the culture we live in and how to navigate its pitfalls. McKee (1997) defines stories as "equipment for living" and refers to movie themes as "controlling ideas," which are "the purest form of a story's meaning, the vision of life the audience members carry away into their lives." Themes can be defined in a number of ways, often in one or several words, such as Love, Death, Justice, Coming of age, and Man vs. Nature. StudioBinder (2020) adds that there are many recurrent themes in modern cinema including God and Philosophy, Survival, Virtue and Values, Love, Society, Good vs. Evil, and Discovering Self-Knowledge.

Themes are often buried deep within a story's structure, making them challenging to articulate or even recognize. However, they serve as powerful vehicles for exploring and communicating ideas, emotions, and experiences through storytelling. In various forms of media, including movies and books, six commonly explored themes emerge: God and Philosophy, Survival, Virtue and Values, Love, Society, Good vs. Evil, and Self-Knowledge. Each theme offers a unique lens through which to examine the human experience.

The theme of God and Philosophy grapples with profound questions such as faith versus doubt, fate versus self-determination, and the tension between reason and faith. It also addresses the existential struggle of facing an empty universe and the potential destruction of the soul. A notable example of this theme is found in Joel and Ethan Coen's film *No Country for Old Men* (Coen & Coen, 2007), which intricately weaves these philosophical dilemmas into its narrative. Transitioning from philosophical inquiries to the primal instinct of Survival, this theme delves into the struggle to endure against overwhelming odds. Films such as *Touching the Void* (Macdonald, 2003), *Life of Pi* (Lee, 2012), *127 Hours* (Boyle, 2010), and *Open Water* (Kentis, 2003) vividly illustrate the human spirit's resilience in the face of adversity (Town & Country, 2022). These stories not only highlight physical survival but also explore the psychological and emotional challenges that accompany such experiences.

As we navigate the complexities of survival, the theme of Virtue and Values emerges, examining the moral and ethical principles that guide human behavior. This theme encompasses topics such as justice, honor, and integrity, prompting audiences to reflect on what it means to live a virtuous life (StudioBinder, 2020). In this context, characters often face dilemmas that test their values, leading to profound personal revelations. Intertwined with virtue is the theme of Love, which explores the multifaceted nature of relationships, including romantic, familial, and platonic love. The challenges and obstacles that arise within these relationships are central to many narratives, inviting audiences to contemplate the complexities of human connection. A quintessential example of this theme is found in the classic film *Casablanca* (Curtiz, 1942), which beautifully captures the nuances of love amid conflict and sacrifice.

Shifting our focus to the broader context of Society, this theme investigates the interactions between individuals and the larger social structures in which they exist. It encompasses social norms, cultural values, and power dynamics, often highlighting contemporary issues. A poignant example is Spike Lee's *Do the Right Thing* (Lee, 1989), which addresses racism and social justice, prompting viewers to reflect on their roles within society. The conflict between Good and Evil is another prominent theme, exploring the eternal struggle between opposing forces. This theme encompasses heroism, redemption, and sacrifice, often portraying characters who embody these ideals. Iconic films such as *Star Wars* (Lucas, 1977), *The Lord of the Rings* (Jackson, 2001–03), and the *Harry Potter* series (Columbus et al., 2001–11) illustrate this theme, captivating audiences with their epic narratives of good triumphing over evil.

Finally, the theme of Self-Knowledge invites audiences on a journey of self-discovery and personal growth. This theme addresses identity, self-awareness, and transformation, often manifesting in coming-of-age stories. A notable example is Amy Heckerling's *Clueless* (Heckerling, 1995), which, inspired by Jane Austen's novel *Emma*, explores the protagonist's journey toward understanding herself and her place in the world. In conclusion, these six themes—God and Philosophy, Survival, Virtue and Values, Love, Society, Good vs. Evil, and Self-Knowledge—serve as foundational elements in storytelling. They not only enrich narratives but also provide audiences with valuable insights into the human experience. By exploring these themes, storytellers can engage viewers and readers in meaningful ways, prompting reflection and discussion about the complexities of life.

Absolutely, in the Moroccan context, a movie theme is the central idea or message of a film, expressed through the characters, plot, and setting. Themes are crucial because they help to give meaning to a film and can make us think about the world in new ways. Moroccan cinema has produced many films with powerful themes, including poverty and inequality, corruption, social injustice, identity and belonging, the role of women in society, the importance of tradition, and the search for meaning in life. These themes are explored in a variety of ways and Moroccan cinema continues to produce films that challenge the status quo and inspire audiences around the world.

In the film analysis below, we organize a number of types of themes into seven major subjects. For each of these subjects, we generate a list of subtopics, then analyze how classic films transform several of those subtopics into common themes in movies. Any given movie, of course, can yield multiple and overlapping movie theme examples from across these major subjects.

Environmental Cinema

The concept of environmental cinema, also known as eco-cinema, refers to the visual representation of the non-human world, particularly nature, and the struggles over its meaning and preservation. This genre of film often focuses on climate and environmental issues as its main theme. These films can rely on the reality and objectivity of scientific facts or serve as a medium for conveying other messages or purposes (Books & ideas, n.d.; Dorlin, 2019; Geal, 2021). Environmental cinema has the power to raise awareness about the biggest environmental problems of our time, including climate change, food waste, water ecosystems, the animal industry, and sustainability (Earth.Org, n.d.). By presenting these issues in an entertaining yet educational format, environmental cinema can inspire viewers to take action towards creating a more sustainable future.

Place of Nature in Cinema

Nature has not always held a place of honor in the seventh art. Its representation has had different uses and meanings over time and across cinematic movements. Bonvoisin (2013) explains that the evolution began with the first shots of the Lumière operators, who carried their heavy equipment around cities as witnesses to a modern and urban era. In this context, nature did not have a significant place, even in the creation of the first studios where it served as a simple backdrop, sometimes even painted in the background. For a long time, the main subject in cinema remained the character, with nature only serving to locate a scene or action.

Environmental cinema is a genre of film that focuses on environmental issues, including climate change, pollution, deforestation, and other concerns. This genre can be used to raise awareness of environmental issues, educate people about the impact of human activities on the environment, and inspire action to protect it. Environmental cinema has gained popularity in recent times as people have become more aware of the threats facing the environment. Some of the most famous environmental films include *An Inconvenient Truth* (Guggenheim, 2006), *The Cove* (Psihoyos, 2009), *Chasing Ice* (Orlowski, 2012), and *The Lorax* (Renaud & Balda, 2012). This genre can be a powerful tool for raising awareness and inspiring action. By watching environmental films, people can learn about the threats facing the environment and the actions they can take to help protect it.

Carbon Footprint of Cinema Industry

Indeed, the film industry's environmental impact is a significant concern. A study by the British Film Institute (BFI) (Hoad, 2020) has called for the film industry to address its environmental impact, including the possible mandatory reporting of carbon emissions. Blockbuster films with budgets over $70 million produce an average of 2,840 tonnes of CO_2 per production, equivalent to the amount absorbed by 3,700 acres of forest in one year (Forbes, 2022). Transport accounts for 51 percent of these emissions. High-definition movies consumed on streaming apps also produce carbon emissions due to the power-intensive and heat-generating data centers that store and deliver the content.

Efforts are being made to reduce the carbon footprint of the film industry. Virtual production technologies are reducing and sometimes eliminating travel associated with film and TV productions. Some productions have implemented recycling programs and sourced materials from sustainable sources. The French National Center for Cinema and Animated Image announced an Action Plan "for a public policy of ecological and energy transition in the cinema, audiovisual and animated image sectors" (Bigmedia, 2022). Eco-production aims to rethink each step of making a film to make it more responsible.

From 2023, each producer will have to provide their carbon footprint. The role of an eco-referent is to reduce the carbon footprint of a film by evaluating the energy expenditure of the shoot and finding solutions (Vallet et al., 2022). The software records all CO_2 expenses and identifies what pollutes the most on the shoot.

The film industry generates around 300 million tons of CO_2 emissions each year, equivalent to the annual emissions of a country like Canada (Goodvest, n.d.). It also produces a large amount of waste, including plastic, paper, and electronic waste. Transportation of films and film crews contributes to pollution due to fuel consumption. The film industry can reduce its environmental impact by using more sustainable materials, investing in renewable energy, reducing waste, using more efficient transportation methods, and offsetting its emissions. By taking these steps, the film industry can help create a more sustainable future.

Approaches to Cinema Analysis

Cinema analysis is the study of film as a work of art, focusing on its formal elements, narrative structure, and social and cultural significance. There are several approaches to cinema analysis, each with its own set of methods and techniques that allow for a deeper understanding of films. One common approach is Formal Elements Analysis, which focuses on the film's formal components, such as lighting, camerawork, editing, and sound. This method can be instrumental in analyzing a film's visual style, narrative structure, and the emotional impact these elements have on the viewer.

Transitioning from formal elements, Textual Analysis involves a close examination of the visual and narrative components of films. For instance, in the context of Moroccan cinema, this approach scrutinizes how the environment is represented, exploring the themes and messages conveyed through these portrayals and the techniques filmmakers employ to communicate them (The Writing Center, n.d.). This analysis allows for a nuanced understanding of how visual storytelling shapes audience perceptions.

Building on textual insights, Historical or Cultural Analysis places the representation of the environment within a broader historical and cultural context. This approach explores how environmental issues have been depicted in Moroccan cinema over time, considering the cultural and social factors that have influenced these representations (Oxford Bibliographies, n.d.). By situating films within their historical milieu, this analysis reveals the evolving relationship between cinema and societal values.

Another significant approach is Narrative Structure Analysis, which focuses on the film's plot, characters, and themes. This method can also delve into the film's intertextual connections, examining its relationship to other forms of art, such as literature, theater, and music (The Writing Center, n.d.). By understanding how narratives are constructed and how they resonate with other artistic expressions, viewers can gain insights into the film's deeper meanings.

Finally, Social and Cultural Significance Analysis emphasizes the film's representation of various social issues, including race, class, gender, and sexuality. This approach explores the film's relationship to the historical and cultural context in which it was produced, allowing for a critical examination of its social relevance and impact (Understanding Media and Culture, n.d.). These diverse approaches to cinema analysis—Formal Elements Analysis, Textual Analysis, Historical or Cultural Analysis, Narrative Structure Analysis, and Social and Cultural Significance Analysis—offer valuable frameworks for understanding the complexities of film. Each method contributes to a richer appreciation of how cinema engages with and reflects the human experience, making it a powerful medium for storytelling and cultural expression.

The method used will be determined by the study topic and aims. Consultation with an adviser or a specialist in the subject can assist in determining the best technique for a research project. Regardless of method, cinema analysis may be a gratifying way to study a film's originality, assisting the viewer in understanding its potential to transmit ideas, entertain, and move them emotionally.

Visual as language Approach

Absolutely, film functions similarly to language, with its own set of grammatical rules and methods of construction. Film experts and creators often describe cinema as a language, with its own vocabulary, symbols, and structure for conveying emotions, ideas, and experiences (Reid & Sanders, 2021). According to Hurbis-Cherrier (2013), the visual language of cinema consists of four main components that work together to create a cohesive narrative: the shot, the sequence of shots, the scene, and the dramatic sequence. These elements serve as the building blocks of cinematic storytelling, allowing filmmakers to craft engaging stories that resonate with audiences.

The most fundamental unit of film language is the Shot, which is a single continuous recording made by a camera. The length of a shot can vary significantly, from a fraction of a second to several minutes, depending on the filmmaker's creative intent and the demands of the narrative (Hurbis-Cherrier, 2013). The shot is the basic unit that filmmakers use to capture and convey visual information to the audience.

Moving beyond the individual shot, the Sequence of Shots is a collection of shots edited together to represent a single action or event, or a change in time or location. By arranging shots in a particular order, filmmakers can create a narrative or convey a specific mood or emotion (Hurbis-Cherrier, 2013). The sequence allows for a more complex and nuanced storytelling approach, as the juxtaposition of shots can create new meanings and associations in the viewer's mind.

At a larger scale, the Scene is a narrative unit that takes place in a single location and deals with a single action. Scenes are used to advance the plot, develop the characters, and create a sense of continuity within the story (Hurbis-Cherrier, 2013). By focusing on a specific location and action, scenes provide a sense of unity and coherence to the narrative, helping to guide the audience through the story.

Finally, the Dramatic Sequence is a series of scenes that form a larger narrative unit, often used to create a more complex narrative structure and engage the audience's emotions (Hurbis-Cherrier, 2013). Dramatic sequences allow filmmakers to explore themes, character arcs, and plot points in greater depth, creating a more immersive and engaging cinematic experience for the viewer.

Finally, Hurbis-Cherrier (2013) adds that a frame, which has both physical and compositional definitions, is the starting point for considering the visual and design aspects of shots. The frame is what the camera's lens is capturing at any given moment, and it's where the director and cinematographer compose the visual elements of the shot. Understanding these components of film language can greatly enhance our appreciation and understanding of cinema.

Wide Shots, Close-Ups, Mid-Shots

In cinema, different types of shots are used to convey different information about characters and their place in the world. It is defined in Hurbis-Cherrier (2013) that the shot, which is the smallest unit of the film language, is "a continuous run of images, unbroken by an edit. Technically speaking, a shot is the footage generated from the moment you turn on the camera to the moment you turn it off—also called a camera take" (p. 43). Wide and medium shots are great for showing off settings and locations; while close-ups and differently angled shots help the audience understand what a character is feeling. Reid and Sanders (2021) assume that wide shots are more functional in nature, serving primarily to establish physical context. Such shots serve to establish a spatial context for an audience and are therefore an important part of most productions. Additionally, mid-shot (typically encompassing a subject from at least the top of their head down to their lower abdomen) helps to provide a broad overview of a person's body language.

The distance between the camera and the subject is critical in this procedure. Wide pictures, for example, may miss subtle facial clues or body language, whereas close-ups are ideal for catching face expressions but may block off body movements entirely. Medium shots strike a balance between the two, capturing an actor's entire physical performance. These different shot types are used in combination to tell a story and convey information to the audience. Conversely, a close-up (which focuses almost all attention on the subject's face and/or eyes) helps to reveal a person's emotional state by laying bare otherwise imperceptible changes in their facial expressions.

The approach, visuals as language, we are using in this chapter underpins that visuals are ways that a film producer uses to communicate information to the audience by carefully considering the camera's relationship to the films subject. For example, Schenk and Long (2021) proposes that by pulling the camera back, the significance of the individual diminishes as they are given less and less on-screen space to occupy. Pulling the camera far enough back, however, can have a devastating impact upon the audience's ability to relate to any person within a frame.

Landscape, Nature and Environment in War Films

Landscape, nature, and environment have long been important elements in war films. They can provide a backdrop for the action, a source of danger or beauty, or a metaphor for the human condition. In some war films, the landscape is simply a setting for the conflict, such as the lush jungles of Vietnam in *Apocalypse Now* (Coppola, 1979) or the deserts of North Africa in *Lawrence of Arabia* (Lean, 1962). In other films, the landscape is more than just a backdrop and can be a source of danger or even a character in its own right, such as the treacherous mountains in *The Thin Red Line* (Malick, 1998) or the unforgiving ocean in *Das Boot* (Petersen, 1981).

The natural world can also be used to symbolize the human condition. In *All Quiet on the Western Front* (Milestone, 1930), the beautiful French countryside is contrasted with the horrors of war, suggesting that war has corrupted both the natural world and the human spirit. In recent years, there has been a growing trend in war films to focus on the environmental impact of war, with films such as *The Hurt Locker* (Bigelow, 2008) and *War Horse* (Spielberg, 2011) showing how war can devastate the natural world.

Rosa (2019) analyzes landscape as a complex cinematic element linked to the representation of corporeality, conquest, and dominance in films and miniseries portraying contemporary warfare in the twenty-first century. Nature and the environment are often used to create realism and heighten emotional impact. For example, in *The Mercy of the Jungle* by Rwandan filmmaker Joel Karekezi, two soldiers lost in the jungle during the Congo War are surrounded by dense foliage and vivid natural sounds that heighten the dizzying nature of war without glorifying pain and suffering (Loayza, 2021). Adam DJ Laity's video essay discusses how filmmakers use landscape, nature, and environment to create new stories within the genre (Miller & Kelly, 2017; Armitstead, 2021; Filmscalpel, n.d.).

Sotto Voce: a Film That Binds both the Environment and War

A film can bind both war and the environment by using the natural setting to enhance the story and convey the emotional impact of war. For example, a film might show how the environment affects the characters, such as soldiers struggling to survive in harsh conditions or being affected by the beauty or destruction of their surroundings. The environment can also be used to heighten the tension and drama of the story, such as a battle taking place in a dense jungle or a desert wasteland. In addition, a film can also explore the impact of war on the environment, such as showing how military operations can cause damage to natural habitats or how conflicts can disrupt local ecosystems. By incorporating these elements into the story, a film can create a powerful and thought-provoking narrative that explores the complex relationship between war and the environment.

The environment can have a significant impact on the outcome of a war film. By using the natural setting to enhance the story, filmmakers can create a more immersive and emotionally engaging experience for the audience. For example, a film that takes place in a harsh or beautiful environment can heighten the tension and drama of the story, making the audience feel more invested in the characters and their struggles. In the same line of thought, the environment can also be used to convey important themes and messages. For example, a film that explores the impact of war on the environment can raise awareness about important issues such as environmental destruction and conservation. By incorporating these elements into the story, filmmakers can create a powerful and thought-provoking narrative that resonates with audiences and leaves a lasting impact.

Overall, the environment can play a crucial role in shaping the outcome of a war film, by enhancing the storytelling, conveying important themes, and creating an immersive and emotionally engaging experience for the audience. In this direction, the Moroccan filmmaker Kamal Kamal perceives himself as an intercultural mediator, as his films allow him to tell stories of the same and the different, the here and the elsewhere, opening up some dialogue and testimonies. Films like *Sotto voce* (Kamal, 2014), *La symphonie marocaine* (Kamal, 2005), *Tayf Nizar* (Kamal, 2002) have the capacity to stimulate the intercultural dialogue. Kamal Kamal's repertoire offers new perspectives on African cinema as it opens up a universal perspective (Serghini, 2018).

Research Hypotheses and Questions

The representation of the environment in Moroccan Cinema is a topic that has not been extensively explored. This research aims to fill this gap by investigating how the environment is portrayed in Moroccan films, and the impact of these representations on society's perception of environmental issues. To achieve this, the following research questions will be addressed:

How has the representation of the environment evolved in Moroccan Cinema over time?

How do Moroccan filmmakers portray the environment in their films?

What is the impact of Moroccan Cinema on society's perception of environmental issues?

These research questions will guide the investigation and provide a framework for analyzing the data collected. By exploring these questions, this research aims to contribute to our understanding of the representation of the environment in Moroccan Cinema and its impact on society.

Methodology

In this study, we employed a visual approach as our theoretical framework to analyze the film, allowing us to examine the visual elements of the film and their significance in conveying meaning and emotions. Our data consisted of a detailed description of the film, including its plot, characters, and visual elements. We carefully analyzed this data to identify the main themes of the film and their significance.

One of the main themes we identified in the film was the environment. We examined how this theme was portrayed in the film and its relationship to different spheres of human life. Our analysis revealed that the environment serves as a catalyst for human emotions, a final resting place, and a source of reconciliation in the film. Overall, our methodology involved using a visual approach to analyze the data and identify the main themes of the film, allowing us to gain a deeper understanding of the significance of the environment in the film and its relationship to human life.

Theoretical Framework: The Visual Approach

The visual approach to film analysis is a critical method that focuses on the visual elements of a film, such as lighting, camerawork, editing, and color. This approach can be used to analyze films from various perspectives, including historical, cultural, and psychological. One key concept in the visual approach is mise-en-scène, which refers to the arrangement of all elements in a film frame. Mise-en-scène can create mood or atmosphere, convey information about characters or the story, or simply create a visually appealing image (University of North Carolina at Chapel Hill, n.d.). Additionally, camerawork is another important concept in the visual approach and refers to how the camera is used to film a scene. Camerawork can create realism, suspense, or simply a visually interesting image (Dinur, 2017; Filmmaker Freedom, 2017). Furthermore, editing is also an important element of the visual approach and refers to how shots in a film are put together. Editing can create rhythm, suspense, or simply a visually interesting sequence (Cinema Punk, 2022). Moreover, color is another important element of the visual approach and can create mood or atmosphere, convey information about characters or the story, or simply create a visually appealing image (Synchronicity Earth, 2020). Overall, the visual approach is a powerful tool for film analysis and can be used to analyze films from various perspectives and reveal hidden meanings and messages (Glaw et al., 2017). A visual analysis can provide valuable insights into how Moroccan filmmakers use visual storytelling techniques to represent environmental issues and convey their messages to the audience.

Data Description

During the Algerian War, Oujda and its surroundings constituted the political hinterland of the Algerian revolution and the refuge of those fleeing the war. The region of Beni Boussaid, on the Moroccan-Algerian border ten kilometers from Oujda, was the crossing point for arms and supplies provided by Morocco and international organizations to the ALN (National Liberation Army). In 1958, to suffocate the ALN inside, French Defense Minister André Morice built an electrified line separating Algeria from the Moroccan borders. The first attempts to cross the diabolical line failed tragically, and the ALN was isolated inside. A solution had to be found to force this passage. Moussa, a Moroccan sympathizer of the revolution, is the smuggler who is charged with passing groups of fugitives through the mountains of Tlemcen (Algeria) to Berkane (Morocco). He arrives in Tlemcen to carry out his usual mission; he finds that this time he will have to pass a group of DHH who come from the Bechar desert. The person in charge on site informs him that the path he usually takes has been sold. He must take the path of Beni Boussaid. Except that the latter is on the Morice line, a belt of 700kilometers along the Moroccan-Algerian borders. It is electrified and full of mines.

Sotto Voce is a film about war. War films set in cold and snowy environments can have a powerful impact on the audience. The harsh conditions of these environments can heighten the tension and drama of the story, making the audience feel more invested in the characters and their struggles. For example, a film set in a snowy battlefield can convey the physical and emotional challenges faced by soldiers fighting in extreme weather conditions. Additionally, the cold and snow can create a sense of isolation and vulnerability, increasing the emotional impact of the story. The film shows characters struggling to survive in a harsh environmental location, highlighting their vulnerability and desperation. It is assumed that war films set in cold and snowy environments can use these conditions to create a more immersive and emotionally engaging experience for the audience by heightening the tension and drama of the story and conveying the physical and emotional challenges faced by the characters. There are many examples of war films set in cold and snowy environments that have been well received by audiences and critics, such as *The Grey* (Carnahan, 2011), *Eight Below* (Marshall, 2006), and *Alive* (Marshall, 1993).

Main Themes of the Film

During the Algerian War, Oujda and its surroundings constituted the political hinterland of the Algerian revolution and the refuge of those fleeing the war. The region of Beni Boussaid, on the Moroccan-Algerian border ten kilometers from Oujda, was the point of passage for arms and supplies offered by Morocco and international organizations to the ALN (National Liberation Army). In 1958, to suffocate the ALN inside, French Defense Minister André Morice built an electrified line (*Ligne Morice*: Morice line) separating Algeria from the Moroccan borders. The first attempts to cross the diabolical line failed tragically, and the ALN was isolated inside. A solution had to be found to force this passage.

Moussa, a Moroccan sympathizer of the revolution, is the smuggler responsible for passing groups of fugitives through the mountains of Tlemcen (Algeria) to Berkane (Morocco). When he arrives in Tlemcen to carry out his usual mission, he finds that this time he will have to pass a group of DHH who come from the desert of Bechar. The person in charge on-site informs him that the path he usually takes has been sold. He must take the path of Beni Boussaid. Except that this one is on the Morice line, a belt of 700 kilometers along the Moroccan-Algerian borders. It is electrified and full of mines.

Kamal Kamal did not live through the Algerian War, of course. But his entire childhood was steeped in the stories his mother told him. His mother was an Algerian woman who herself was sentenced to death in absentia and had to make this dangerous crossing from Tlemcen to Oujda—a crossing that her son later on filmed with great narrative mastery in *Sotto Voce*.

Results

In the film *Sotto Voce*, the environment plays a significant role in shaping the experiences and emotions of the characters. It serves as a catalyst, triggering human feelings and emotions. For instance, the portrayal of the environment is linked to the DHH characters in the film. These characters are often shown in nature, highlighting its beauty and significance.

The environment is also connected to the cycle of life and death in the film. Individuals nearing the end of their lives are often depicted in nature, suggesting a connection between the environment and the end of life. This portrayal of nature as a final resting place underscores its significance in human life.

Despite its prominence in the film, the environment is not explicitly portrayed as a source of environmental awareness or activism. The film does not provide any explicit environmental awareness tips or messages. However, nature is portrayed as a source of reconciliation in the film. Characters are often shown finding peace and solace in nature, suggesting its restorative and healing properties. Overall, the environment in the film *Sotto Voce* is intricately connected to different spheres of human life. It serves as a catalyst for human emotions, a final resting place, and a source of reconciliation.

Shots of different types are an important tool for filmmakers to convey the theme of their film, especially when the environment is a background theme. By focusing on the details of a character's face or actions, close shots, for example, can reveal their emotions and reactions to the world around them. This can help to create a deeper connection between the audience and the characters, allowing viewers to better understand their motivations and experiences. Additionally, shots, whether they are wide shots, close-ups, or mid-shots, can be used to highlight specific elements of the environment, drawing attention to their significance within the story. Overall, the use of various shots can add depth and nuance to a film's portrayal of its themes.

Title of the Film

"Sotto voce," an Italian phrase that could be translated to "under voice" or "in a low voice," is often used in music to indicate that a passage should be played or sung softly or in a subdued manner. In a broader context, it can refer to speaking or expressing oneself in a quiet or discreet manner. The title "Sotto Voce" could indeed symbolically relate to the film's themes and narrative in several ways. The meaning of speaking softly or discreetly could represent the characters' attempts to navigate challenging situations while avoiding detection. The idea of reducing other voices to silence could also reflect the suppression of dissenting voices during times of conflict. The portrayal of the natural world in the film, serving as a backdrop for the story, could represent the larger context in which the characters' struggles take place.

Opening and Closing Scenes: From Nature to Urbanism Dawn

The opening and closing scenes of a film are important elements of storytelling. The opening scene (Figure 9.1) sets the tone for the rest of the film and introduces the audience to the characters, setting, and plot. The closing scene, on the other hand, wraps up the story and provides closure for the audience. The film is set during the Algerian war and follows the story of a Moroccan friend of the revolution who helps refugees across the mountains. Generally, filmmakers use various techniques to convey their message and set the tone for their films. Opening a war film with scenes of nature, as it is the case with *Sotto*

Figure 9.1 Opening scene: Environment. Source: Kamal Eddine Benabid.

Voce, could serve several purposes. The scenes are used to contrast the beauty and tranquillity of nature with the violence and chaos of war. Furthermore, they are used to establish the setting and to provide context for the story.

Based on our observation, the image depicts a natural setting during autumn, with a long stretch of barbed wire dividing the scene. The fence represents a boundary or barrier, separating two different areas or territories. The contrast between the natural beauty of the autumn landscape and the harshness of the barbed wire conveys a sense of tension and conflict.

The time of the scene is dawn, a lapse that marks the beginning of twilight before sunrise. The faint glow of twilight signaled the start of another day on the front lines. In films about nature, dawn can symbolize illumination and hope, the beginning of a new day and thus a chance for happiness and improvement, a symbol of birth and rebirth, and a sign of awakening. The coming of light represents resurrection, transformation, and renewed passion. So, dawn in this film is an important theme to start with as it represents new beginnings and hope. However, dawn on a gloomy and a snowy morning has a different atmosphere compared to a clear or sunny day. The overcast sky can create a subdued and moody ambiance. The lack of direct sunlight can make the colors of the landscape appear more muted, adding to the melancholic feeling. Still, dawn in a bloomy day represents the beginning of a new day and the potential for change and improvement. The soft light filtering through the harsh clouds can provide a sense of calm and tranquillity.

Time and Place Shift

In the context of the film *Sotto Voce*, a shift in time from 1958 to the twenty-first century and a shift in place from the border between Algeria and Morocco could represent significant changes in the characters' lives and circumstances. The shift in time could show how the characters have aged and how their experiences during the Algerian war have shaped their lives. The shift in place represents a change in the characters' environment and the challenges they face as they move from one location to another. These shifts could also be used to explore broader themes such as the passage of time, the impact of historical events on individuals, and the relationship between people and the natural environment.

In the opening scene of the film, close and wide shots of decay are used as a symbol to represent decline and decomposition, an image of moral decay. The

Figure 9.2 Opening scene: Decay. Source: Kamal Eddine Benabid.

end of the world in this narrative is so severe because of the loss of nature. In the opening scene of the film, set in 2013, the camera pans over a desolate landscape. The shot focuses on three people—two old men and a woman—squatting amid decaying machinery (Figure 9.2). This shot represents the modern world, where industrialization and progress have left their mark. But as the scene progresses, we see another group of people attempting to flee across the Algerian border into Morocco. These characters are always shown in relation to nature, highlighting their vulnerability and desperation as they make their perilous journey. The contrast between these two shots—the decaying machinery and the natural landscape—serves to underscore the stark differences between the world of progress and the world of survival. The film uses this powerful imagery to draw attention to the ongoing struggles faced by those living on the margins of society.

Catalyst Event

In a movie, the catalyst is the event that takes place that causes the protagonist to take action (Lim, 2012). In addition to the Big Event, Trottier (2014) assumes that there are six other key plot points that are essential to a successful movie. Together, these seven plot points make up what he calls "TheMagnificent7 Plot Points" that include the "Back Story," the "Catalyst," the "Midpoint," the "Crisis," the "Climax," and the "Realization." Usually, a catalyst event happens within the first ten minutes of the movie (Lim, 2012). The catalyst is what kicks things out of balance and gives the protagonist a new problem or desire. It is often defined with regard to the opening or starting of a film (Lim, 2012) or the inciting incident that hooks the viewer into the story and sets everything else that happens into motion (StudioBinder, 2019).

In the film *Sotto Voce*, the inclusion of a dramatic event involving an aggrieved girl who is on the verge of killing herself could serve as a catalyst for bringing together the two separate dimensions of time and place (Figure 9.3). The intervention of an old man who tells her the story of the film and brings it back to the colonial period could provide a link between the past and the present and between the characters' experiences during the Algerian war and

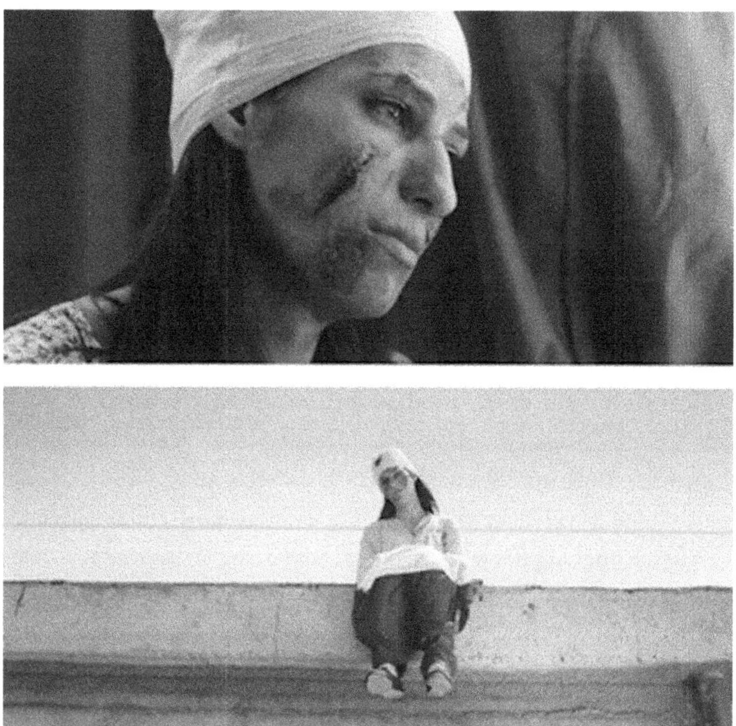

Figure 9.3 Aggrieved girl. Source: Kamal Eddine Benabid.

their lives in the twenty-first century. This narrative device could be used to explore themes such as the impact of trauma, the power of storytelling, and the connections between different generations.

Nature in the Film is linked to the DHH Characters

In terms of nature, the film adopts the environment to have a healing effect on people's suffering. Being in nature, or even viewing scenes of nature, reduces anger, fear, and stress, and increases pleasant feelings. For instance, exposing the DHH characters to nature (Figure 9.5) not only improves their emotional well-being but also contributes to their physical health. As for the unbearable suffering, it is alleviated with nature throughout this film, which took nearly three years to shoot. Fulfilling his passion for music, Kamal incorporates his own opera compositions within the storyline. These compositions act as a

Figure 9.4 DHH characters. Source: Kamal Eddine Benabid.

Figure 9.5 DHH characters and nature. Source: Kamal Eddine Benabid.

juxtaposition to the brutal reality of war and serve as a strong and recurring motif throughout the film (Joucka, 2015).

In the film, the DHH characters are described as "choir singers who hear nothing and listen only to their inner voice." We can assume that they are the only pure and innocent individuals who are not tainted by all the different ideologies being spread around at that time—simply because they are not able to hear them. This portrayal of the DHH characters as pure and innocent individuals who are not influenced by the ideologies of the time highlights their unique position in society. By not being able to hear, they are not exposed to the same influences as others and are able to maintain their innocence and purity.

The film *Sotto Voce* can be seen as not only a film about war, but also as a film that explores the depiction and imagination of borders in film. While films have explored national and political borders, this one attempt to identify, challenge, and imagine frontiers of another kind: social, ethnic, religious, and gendered. The film addresses a problem that occurred in the 1950s, when the separation between countries was established by the colonizer. This issue is evoked and linked to its current context, as the borders between Morocco and Algeria have been closed due to tensions between the two countries since 1994.

The closure of the borders has had a significant impact on the lives of families separated by the border, as well as on the economic and political integration of the region. Despite the historical roots of this issue, it remains relevant today and continues to affect the relationship between these two neighboring countries. The closure of the borders between Morocco and Algeria since 1994 has had an impact on families separated by the border. Moroccan and Algerian families are torn apart, or even impoverished, by the closed border between the two countries. It should also be noted that the closure of the borders between Morocco and Algeria is one of the main factors not only preventing the emergence of the region on a political level but also blocking any economic and financial integration between the countries of the region.

This unique focus on the representation of borders and frontiers via film is groundbreaking, opening up a new field of research and scholarly discussion. The exceptional variety of national and cultural perspectives provides a rewarding investigation of borders and frontiers. While humans have the ability to establish borders and close frontiers among themselves, nature is a free extension that does not acknowledge these restrictions. The natural world exists without regard for the political boundaries that humans have created, and its processes and movements continue regardless of the barriers that we erect. Rivers flow, animals migrate, and weather patterns shift without regard for the lines on a map. This serves as a reminder that, while we may try to impose our will on the world around us, nature ultimately operates according to its own rules and rhythms.

When nature is presented as an appreciated element in a film, it is often done through the DHH characters. It is through their senses and feelings that we, as the audience, are brought to appreciate nature. Even if the characters are DHH, the unspoken voice of the natural setting can still be heard, felt, and smelt. In *Sotto Voce*, which is a war film, the focus is on the conflict and violence between human characters. The narratives on gunfire, explosions, and shouting dominate the audio landscape, drowning out the more subtle sounds of nature. In this way, the voice of nature is silenced, as the natural world becomes a mere backdrop to the human drama unfolding on screen. This silencing of nature can be seen as a metaphor for the destructive impact of war on the environment. As human conflicts rage on, the natural world is often caught in the crossfire, with habitats destroyed and wildlife displaced. The presence of nature's voice in war films serves as a reminder of the toll that war takes on the environment.

Nature as a Trigger of Human Feelings

Nature has long inspired people to love, appreciate beauty, and express their artistic abilities. The natural world, with its diverse landscapes, vibrant colors, and intricate patterns, provides a rich canvas for artistic expression. From the delicate petals of a flower to the majestic sweep of a mountain range, nature offers endless possibilities for creative exploration.

In addition to inspiring artistic expression, nature also has the power to bring people together and foster love and connection. The beauty and tranquillity of the natural world can provide a peaceful setting for people to connect with each other on a deeper level. Whether it's watching a sunset together or taking a walk in the woods, nature provides opportunities for people to share experiences and bond over their appreciation of its beauty (Figure 9.6).

Figure 9.6 Humans in nature: A place for all. Source: Kamal Eddine Benabid.

Of great importance to human wellbeing, nature is a powerful force that inspires people to love, make them appreciate beauty, and help them express their artistic skills. By connecting with the natural world, we can tap into its endless well of inspiration and find new ways to express ourselves and connect with others. The natural setting plays a role in the story as Moussa helps refugees navigate the dangerous landscape. Nature provides a wealth of gifts to humans, including physical sustenance, health and healing, beauty and inspiration, connection and community, and spiritual growth. Spending time in nature has been shown to have numerous health benefits, including reducing stress and improving mood (Figure 9.7).

The natural world is full of beauty and wonder that can inspire us to create art and other forms of creative expression. Nature also provides opportunities for people to come together and connect with each other. For many people, nature is a source of spiritual nourishment and growth.

Research reveals that environments can increase or reduce our stress, which in turn impacts our bodies. Being in nature, or even viewing scenes of nature, reduces anger, fear, and stress and increases pleasant feelings (University of

Figure 9.7 Improving mood. Source: Kamal Eddine Benabid.

Minnesota, n.d.). Spending time in nature is linked to both cognitive benefits and improvements in mood, mental health, and emotional well-being (Mental Health Organization, n.d.). People who feel more connected to nature have greater eudaemonic well-being, a type of contentment that goes beyond just happiness (American Psychological Association, 2020).

In a study, people who took a 50-minute walk in a natural environment experienced decreased anxiety, rumination, and negative affect, as well as preservation of positive affect, and increased working memory performance compared to those who walked in an urban environment (Jimenez et al., 2021). Nature-based health interventions (NBI) are designed to engage people in nature-based experiences with the goal of improving health and wellness outcomes. These interventions facilitate engagement with nature and help individuals overcome mental fatigue and restore their capacity to direct attention (Nejade et al., 2022).

Visuals that Speak

In the realm of film, visuals often substitute for language, conveying meaning without the necessity of words. This is especially true in the depiction of the environment and the natural world. The filmmaker communicates information about the environment to the audience in a potent and effective manner through visual storytelling techniques such as composition, lighting, color, and movement. For instance, the filmmaker employs sweeping shots of a landscape to express the beauty and grandeur of the natural world. The lighting and color in these shots are manipulated to evoke a specific mood and atmosphere, such as the crisp light of dawn or the cool blues of a mountain stream. Likewise, close-up shots of flora, fauna, birds, vegetables, and insects are utilized to highlight the intricate details and patterns inherent in nature. In this manner, visuals act as a form of language, conveying information about the environment (Figure 9.8). By employing visuals in innovative and effective ways, filmmakers can facilitate audience engagement with the natural world, fostering a deeper appreciation for its beauty and complexity.

Death and Nature

In the context of film, visuals can convey meaning without the use of words, particularly when depicting the environment and the natural world. Through

Figure 9.8 Species in nature. Source: Kamal Eddine Benabid.

composition, lighting, color, and movement, filmmakers can communicate information about the environment to the audience. For instance, sweeping shots of a landscape can convey the beauty and majesty of the natural world, while lighting and color create a specific mood or atmosphere. Thus, visuals serve as a language for communicating information about the environment, helping audiences connect with the natural world and appreciate its beauty and complexity.

Films may depict death in nature for various reasons, such as conveying the harsh realities of life and death or representing the end of a journey. In *Sotto Voce*, this imagery highlights themes of loss, memory, and the passage of time. The film explores characters haunted by their past and struggling with their present. By depicting death in nature, the film suggests that our lives are subject to inevitable cycles of birth, growth, and decay. This serves as a poignant reminder of our mortality and the transience of human existence (Figure 9.9).

The relationship between death, snow, and the color white in *Sotto Voce* is complex and multifaceted, with various symbolic meanings depending on the cultural context. In some cultures, white is associated with purity, innocence, and peace, while snow can symbolize a blank slate or a new beginning. In other cultures, white is associated with mourning and grief, while snow can symbolize the finality of death.

The association of death with snow and the color white can have various symbolic meanings depending on the cultural context. In some cultures, white

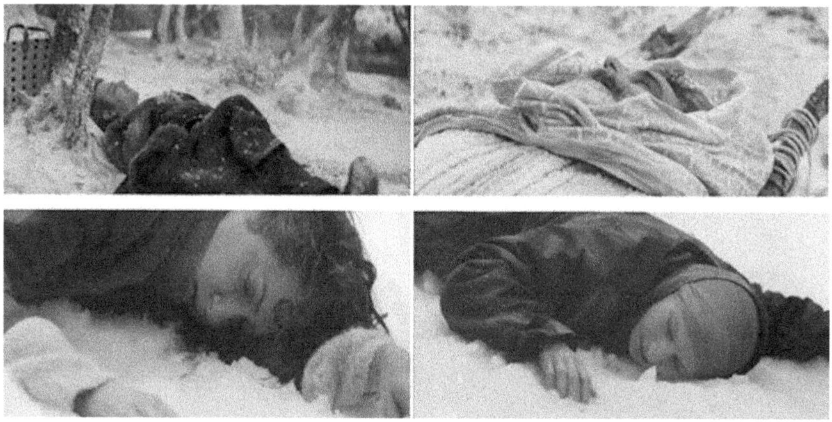

Figure 9.9 Death in nature. Source: Kamal Eddine Benabid.

is associated with purity, innocence, and peace, and may represent the release of the soul from the body or the transition to the afterlife. Snow, with its ability to cover and transform the landscape, can symbolize a blank slate or a new beginning, representing the end of one phase of life and the start of another.

In other cultures, white is associated with mourning and grief, and may represent the emptiness and loss that come with death. Snow, with its coldness and stillness, can symbolize the finality of death and the cessation of life. The specific symbolism of associating death with snow and the color white can vary depending on the cultural context and the themes being explored in a particular work of art or literature.

The absence of certain colors, such as red and yellow, in some films can be a deliberate artistic choice made by the filmmaker. Color can have a powerful impact on the mood and atmosphere of a film, and the use or absence of certain colors can convey specific emotions or ideas. For example, a filmmaker may choose to exclude bright, warm colors like red and yellow to create a bleak or melancholic mood, or to convey a sense of coldness or detachment. In other words, the absence of certain colors may also be used to create visual contrast or to draw attention to specific elements within the frame. For example, a filmmaker may choose to use a limited color palette to make certain objects or characters stand out against the background. The specific reasons for excluding certain colors from a film can vary depending on the artistic vision of the filmmaker and the themes being explored in the story.

In the film *Sotto Voce*, the presence of gloomy colors such as black, white, and gray may have a symbolic meaning. These colors can reflect the unknown fate of the characters who flee execution and death. The use of a muted color palette can convey a sense of uncertainty and fear, as the characters navigate a dangerous and unpredictable world. By using these colors to create a bleak and oppressive atmosphere, the filmmaker may be suggesting that the characters are facing an uncertain future, with the threat of death looming over them at all times.

In the film *Sotto Voce*, the filmmaker's selective use of color during the segments set in the 1950s appears to be a conscious artistic decision. By restricting the color palette to hues that evoke a wintry ambiance, the filmmaker effectively immerses the audience in the specific temporal setting. The characters' attire, largely devoid of vibrant colors, further reinforces this effect. Notable exceptions occur when the color red is employed to accentuate a tomato amidst the snow or to depict blood, thereby creating striking visual contrasts.

Interestingly, the film's transition to the contemporary era is marked by a shift to more chromatically diverse scenes, with objects rendered in their natural colors. This juxtaposition of the past's muted tones with the present's vivid hues serves to underscore the passage of time and the ensuing transformations. Through this nuanced use of color, the filmmaker subtly yet powerfully communicates key aspects of the narrative and its underlying themes. Thus, color becomes an instrumental storytelling device, enhancing the film's overall impact.

No Environmental Awareness Tips

Cinema possesses a potent capacity to heighten environmental consciousness, educating and inspiring audiences about the imperative of environmental preservation. Through compelling narratives and evocative imagery, filmmakers can portray the splendor and vulnerability of the natural world, as well as the consequences of human activities on the environment, thereby advocating for positive change.

In the film *Sotto Voce*, environmental awareness is depicted as a nascent concern. This is exemplified in a scene where Moussa, after lighting a cigarette, casually discards the matchstick into the environment. This act is repeated when he lights a second cigarette for another character, again carelessly tossing the spent matchstick onto the ground (Figure 9.10). Sbaissi, one of the group's

Figure 9.10 Environmental awareness tips. Source: Kamal Eddine Benabid.

leaders, reacts strongly to these actions. However, his protest is not rooted in environmental considerations, but rather in the fear that the smoke could alert the French military to their presence. Moussa dismisses this concern, asserting that the scent of smoke would not be detectable from fifty kilometers away.

This exchange underscores the divergent priorities and concerns of the characters, potentially mirroring the attitudes and priorities of the era in which the film is set. It subtly highlights the lack of environmental consciousness, with the characters' immediate survival taking precedence over long-term environmental considerations. Thus, through such nuanced depictions, cinema can serve as a powerful medium for fostering environmental awareness and promoting sustainable practices.

Screaming in the Environmental Setting

In cinema, screaming serves as a potent narrative device, conveying a spectrum of emotions and significantly enhancing the storytelling process. When set

Figure 9.11 Screaming in a silent setting. Source: Kamal Eddine Benabid.

against the backdrop of open nature or forests, screaming can have a cathartic effect, enabling individuals to vent suppressed emotions and tension.

In *Sotto Voce*, a war film, screaming assumes a symbolic role, representing the physical and emotional distress, fear, and trauma endured by civilians during colonization. It also encapsulates the pandemonium, disarray, and desperation characteristic of war.

The introduction of screaming in an otherwise silent setting can be particularly impactful. The stark contrast between the prevailing silence and the abrupt, loud noise can evoke shock and surprise, intensifying the emotional resonance of the scene. The sudden screams disrupt the tranquillity, instilling a sense of urgency, fear, or desperation.

The interplay between silence and screaming in *Sotto Voce*, underscores the tension and peril confronting the characters as they endeavor to cross the border undetected by the French colonizers. The film's title, translating to "in a soft voice" or "whisper," implies the necessity for stealth and caution. Yet, the presence of screams contravenes this objective, introducing an element of disruption and risk. These screams serve as a constant reminder of the high stakes involved in their mission and the potential repercussions of detection. Moreover, the forest setting can amplify the screams, attracting unwanted attention and escalating the danger (Figure 9.11).

The film employs screams as a powerful protest against execution, liberty, and the eradication of borders. When set against the serene and silent backdrop of the forest, these screams are amplified and extended. The juxtaposition of the peaceful natural setting with the intense human emotions serves to magnify the impact of the screams, drawing attention to the characters' struggles and aspirations. By contrasting the tranquillity of nature with the violence of human conflict, the filmmaker perhaps suggests that amidst chaos and turmoil, hope for a brighter future persists.

Nature as a Source of Reconciliation

Nature can serve as a potent antidote to depression and mental health issues. Immersion in natural environments has been shown to alleviate mental health disorders such as anxiety and depression. Exposure to natural light can be particularly beneficial for individuals suffering from Seasonal Affective Disorder, a type of depression that manifests during specific seasons or times of the year. Elements of nature, such as birds, can inspire individuals to appreciate life, fostering hope by establishing a connection with the natural world and its inherent beauty. This connection can significantly enhance mental health and well-being (Image up left in Figure 9.12).

In the film, a girl who has endured severe aggression draws inspiration from a small bird to emerge from a paper box, a metaphor for her current state of suffering. This transformative moment symbolizes the character's resolve to

Figure 9.12 Birds as a source of reconciliation. Source: Kamal Eddine Benabid.

transcend her pain and suffering and to reclaim her life. The imagery of the character emerging from the box, akin to a bird taking flight, suggests that she has mustered the strength and courage to surmount her challenges and to continue her journey (image up and down left Figure 9.12).

The female character's decision to abandon her suicidal thoughts (Image up right Figure 9.12) and to embrace life represents a powerful testament to hope and resilience. It demonstrates that even in the bleakest of times, there exists the potential for change and growth. Drawing inspiration from the natural world, the character discovers a renewed sense of purpose and meaning, embarking on her journey towards healing and fulfillment (Image downright Figure 9.12).

Closing Scene: From Nature to Urbanism

Dusk, the period following sunset, carries profound symbolism in films. Within an urban context, dusk signifies the transition from the day's hustle and bustle to the calm and tranquillity of the night. The diminishing light can evoke a sense of mystery and anticipation as the city metamorphoses from a hub of activity to a haven of rest and relaxation.

In the film under examination, the depiction of an urban setting at dusk symbolizes the closure of one chapter and the commencement of another. The setting sun signifies the conclusion of a narrative or the resolution of a conflict, while the impending night heralds new possibilities and opportunities. The contrast between the city's artificial lights and the natural light of the sunset creates a visually arresting image, underscoring the interplay between nature and civilization.

In *Sotto Voce*, the portrayal of an urban setting at dusk conveys a sense of transition, transformation, and potential. The concluding scene, which depicts a city at night during wartime (Image up right Figure 9.13), signifies the end of one phase of the battle and the onset of another. It implies that the aggrieved girl must continue her journey in musical apprehension, and that the divided countries must seriously contemplate the eradication of barriers and frontiers.

The transition from a scene set in nature to a dense urban setting carries potent symbolism in films. Nature often symbolizes tranquillity, peace, and harmony, while urban settings can represent chaos, noise, and conflict. The contrast between the two can accentuate the disparities between the natural world and human civilization.

Broadly speaking, this evocative imagery serves as a call to action, urging the characters and the audience to reflect on the past and strive for a better future.

Figure 9.13 Closing scenes. Source: Kamal Eddine Benabid.

The setting sun signifies the conclusion of a battle or the resolution of a crisis, while the approaching night represents new challenges and dangers. Dusk, the period just after sunset, symbolizes the transition from the chaos and violence of war (colonial period) and violence (aggressed girl) to a moment of relative calm and reflection. The fading light engenders a sense of uncertainty and apprehension as the city transforms from a battlefield to a place of relative safety.

In relation to the presence of the environment in the concluding scene, the transition symbolizes a shift in tone and mood. Moving from nature to urbanism represents a change from calm to chaos, or from harmony to conflict. It also signifies a change in perspective, as the film transitions from the simplicity of the natural world (the fifties) to the complexity of human society (real time). The transition from a scene set in nature to a dense urban setting conveys a sense of contrast, change, and tension.

The symbolism of the old man walking away from the audience in the film represents the character's desire to leave his past behind, distance himself from a difficult situation, and move forward to face new challenges. The natural setting adds additional layers of meaning, representing a source of solace and refuge, a challenge and obstacle that highlight the character's isolation and vulnerability. The specific symbolism of this image (Image up left Figure 9.13) depends on the artistic vision of the filmmaker and the narrative being told.

Recycling as a Positive Impact

Recycling is an effective way for individuals to positively impact the environment by contributing to waste reduction and environmental protection. The human connection to nature has shifted from entertainment and appreciation to a preoccupation with preservation due to the realization that human lifestyles have inflicted considerable harm on nature. A movement towards conservation and preservation has grown, acknowledging that our actions have consequences and that we should take responsibility for the harm humans have caused. Recycling saves energy and natural resources, reduces pollution and greenhouse gases, and conserves valuable resources. Recycling programs are managed at the state and local level, and the US Environmental Protection Agency is working to build an economy that keeps materials, products, and services in circulation for as long as possible.

In this new paradigm, the film *Sotto Voce* centers around the idea that nature should no longer be viewed merely as a source of pleasure and entertainment, but as a domain to be protected and preserved. An example of this shift in the relationship between humans and nature can be seen in the sculptures made from recycled materials (iron) displayed on the Casablanca beach, which depict human forms (Figure 9.14). These sculptures reflect the current environmental situation, where recycling harmful materials has been artistically repurposed as a means of raising awareness. This contrasts sharply with the portrayal of nature in the film, which discusses the passage of actors through borders across a gloomy and snowy forest separating Morocco from Algeria.

In *Sotto Voce*, nature is presented as a backdrop to human drama, something to be traversed and overcome. Conversely, the sculptures represent a new way of thinking about our relationship with nature. They demonstrate that we are not separate from nature, but intimately connected to it. It should be recognized that visuals have the power to convey complex messages in an immediate and impactful way. They can communicate ideas and emotions that transcend language and cultural barriers, making them an incredibly effective tool for raising awareness about important issues.

The symbolism of sculptures made from recycled metals in the shape of human beings varies depending on the context and themes being explored. Firstly, these sculptures may represent the idea of transformation and renewal, as discarded materials are given new life and purpose. The use of recycled metals may also suggest a connection between human beings and the environment, highlighting the importance of sustainability and responsible resource use. Furthermore, the human forms of the sculptures serve to personalize and humanize environmental issues, reminding us that our actions have a direct impact on the world around us. By creating art from recycled materials, the artist may be encouraging us to

Figure 9.14 Recycling as a positive impact Source: Kamal Eddine Benabid.

reflect on our relationship with the environment and to consider the role that we can play in protecting our planet for future generations.

Secondly, the public display of human bodies in the form of sculptures made from recycled metals serves as a powerful symbol of humanity's responsibility towards the environment. These sculptures represent a kind of trial in the court of life, where humanity is called to account for its actions and their impact on the natural world. By using recycled materials to create these sculptures, the artist may be suggesting that we all have a role to play in mitigating the environmental impact of our actions. This is a powerful reminder of the need for environmental stewardship and the importance of sustainable practices.

Conclusion

Cinema has always accompanied social, economic, and technological developments. In relation to the environment, cinema has the potential to be a powerful force for good in the world, raising awareness of environmental issues, inspiring people to take action, and promoting sustainable practices. Equally important, films can educate the public about climate change, the destruction of marine life, and the melting of the polar ice caps. Additionally, the cinema can also inspire people to reduce their carbon footprints, recycle and compost, and live more sustainably. In addition to highlighting the negative impact of human actions on the planet, cinema can also showcase the benefits of renewable energy and sustainable agriculture. By showing people how they can make a difference, films can help to create a more sustainable future. Filmmakers have a responsibility to use their platform to raise awareness of environmental issues and promote sustainable practices. They can do this by making films that educate the public, inspire people to take action, and promote sustainable lifestyles.

In this sense, digital cinema allows for ecological advances. However, the environmental concern of our century will not be solved in twenty-five frames per second. The main thing is to inform so that the population can make informed choices and become aware of the problems. The idea is to make the environment a central subject in the media to reach more viewers and raise awareness of the environmental crisis. Making ecological films is good, but making them while respecting the environment is better. The smallest possible step is necessary to move towards a sustainable society.

In the same line of thought, documentaries are one way that filmmakers can propose a certain vision of the world and potentially commit to a cause. The editor's work can balance between technical skills and creative aspiration depending on the space left by his director. Starting his professional life in audiovisuals, working in an ecological documentary is a first solution for an editor to take part in the collective effort. Most agree that these films form a momentum for mobilization. Currently, it is a challenging task for filmmakers to be environmentally conscious while producing environmental films. The limited availability of ethically sound films and the pressure to secure work make it difficult to make the right choices.

In conclusion, cinema has the power to expose environmental issues and raise awareness about the impact of human actions on the planet. Through powerful visuals and storytelling, filmmakers can bring attention to pressing environmental concerns and inspire audiences to take action. While cinema is often associated with fun and entertainment, it can also serve as a platform for social and political commentary. Films that tackle environmental issues can challenge viewers to think critically about their own behaviors and the consequences of their actions. By presenting complex environmental problems in an accessible and engaging format, cinema can help educate and mobilize audiences around important issues. In recent years, there has been a growing trend towards environmental-themed films that seek to raise awareness about climate change, pollution, deforestation, and other pressing concerns. These films often feature stunning cinematography that captures the beauty and fragility of the natural world while also highlighting the devastating impact of human activities.

Recommendations

In summary, our study highlights the importance of environmental themes in cinema and offers actionable recommendations for both researchers and producers. By exploring the role of the environment in film and integrating environmental awareness messages, stakeholders can contribute to a more informed and engaged audience, ultimately fostering a greater appreciation for the environment through the art of storytelling.

For future researchers, we advocate for a deeper exploration of the role of the environment in film and its influence on human emotions and experiences. This could involve studies examining the efficacy of environmental awareness messages in film and the potential of cinema to shape environmental attitudes and behaviors. By investigating these aspects, researchers can contribute valuable insights into how films can serve as tools for promoting environmental consciousness.

In addition to recommendations for researchers, we suggest that cinema producers contemplate the integration of explicit environmental awareness messages in their films. Our study indicates that the environment is a recurrent theme in film and can act as a catalyst for human emotions. By incorporating environmental awareness messages, producers have the opportunity to raise awareness and foster positive environmental attitudes and behaviors among viewers. This approach not only enhances the narrative depth of films but also aligns with a growing audience interest in sustainability and environmental issues.

Furthermore, we recommend that cinema producers consider depicting nature as a source of reconciliation in their films. Our study reveals that nature can possess restorative and healing properties, and its inclusion in films could help cultivate positive attitudes toward the environment. By showcasing nature's beauty and its potential for healing, filmmakers can inspire audiences to appreciate and protect the natural world.

Disclosure

All cinematic photos extracted from the film *Sotto Voce* have been used with the explicit authorization of the film's director, Kamal Kamal. The details of this authorization are provided in Appendix 1 and Appendix 2.

Appendix 1. Request to Mr. Kamal Kamal, director of the film *Sotto Voce*, to obtain authorization to use footage from his film in this chapter.

Docusign Envelope ID: D73B542F-8623-403E-9CBC-E5D9F66546AA

À l'attention de Monsieur Kamal Kamal

Objet: Demande d'autorisation pour l'utilisation d'images du film "Sotto Voce" à des fins académiques

Le 12/08/2024

Monsieur Kamal Kamal,

Nous espérons que cette lettre vous trouve en bonne santé et de bonne humeur. Nous sommes les auteurs d'une étude intitulée "Visual Language and Environmental Representation in Moroccan Cinema: '*Sotto Voce*' as a Case Study", menée par Dr. Mohamed Mliless, Dr. Latifa Belfakir, et Dr. Fouad Souiba.

Dans le cadre de notre recherche, nous souhaitons analyser et illustrer notre étude avec des images extraites de votre film "Sotto Voce". Nous croyons fermement que votre œuvre apportera une valeur significative à notre analyse académique et permettra une meilleure compréhension des thèmes abordés.

Nous vous serions extrêmement reconnaissants de bien vouloir nous accorder l'autorisation d'utiliser ces images à des fins strictement académiques. Nous nous engageons à mentionner explicitement votre nom et celui de votre film dans toute la publication.

Nous restons à votre disposition pour toute information complémentaire et espérons recevoir une réponse favorable de votre part.

Veuillez agréer, Monsieur, l'expression de nos salutations distinguées.

Dr. Mohamed Mliless

Signed by:

Mohamed Mliless

5DE88B35C3F84CC

Appendix 2. Authorization from Kamal Kamal to use footage from his film *Sotto Voce* in this chapter

Benabid Kamal Eddine
2, Rue Mohamed Abdou,
Q. Palmier, Casablanca
benabidkamal4@gmail.com
00212665716633

Objet : Autorisation pour l'utilisation de captures photographiques du film "Sotto Voce" à des fins académiques

Cher Monsieur Mliless,

En vous remerciant pour l'intérêt que vous portez à mon œuvre et à ma personne, je vous autorise à publier les captures photographiques de mon film "Sotto Voce" dans votre publication académique.
I authorize you to publish the photographic captures of my film "Sotto Voce" in your academic publication.

Cordialement

I authorise you To publish the photographic captures of my film "Sotto Voce" in your academic publication

Kamal Eddine Benabid
(Kamal Kamal)

References

Abu Wael, T. (Director. (2001). يوميات عاهر [Diary of a male whore]. [Moving Picture]. Palestine.

Allouache, M. (Director). (1994). *Bab El-Oued city*. [Moving Picture]. Algeria. Blaqout Prod.

American Psychological Association. (2020, April 1). Nurtured by nature. *Monitor on Psychology*, *51*(3). https://www.apa.org/monitor/2020/04/nurtured-nature

Anderson, W. H. U. (Ed.). (2022). *Film, philosophy and religion*. Vernon Press.

Armitstead, C. (2021, June 26). *Stories to save the world: the new wave of climate fiction*. Retrieved from: https://www.theguardian.com/books/2021/jun/26/stories-to-save-the-world-the-new-wave-of-climate-fiction

Ayouch, N. (Director). (2003). *Une minute de soleil en moins* [A minute of sun less]. [Moving Picture]. Arte (France); BC Films (France); Alin' Productions (Maroc); GMT Productions.

Bachir-Chouikh, Y. (Director). (2002). *Rachida*. [Moving Picture]. Algeria: Ciel Production

Bigelow, K. (Director). (2008). *The Hurt Locker* [Moving Picture]. Summit Entertainment.

Bigmedia. (2022). Eco-production: comment le cinémas'adapte aux enjeux climatiques [Eco-production: How cinema adapts to climate issues]. Retrieved from: https://bigmedia.bpifrance.fr/decryptages/eco-production-comment-le-cinema-sadapte-aux-enjeux-climatiques

Bonvoisin, D. (2013). Le cinéma à la conquête des paysages de la nature [Cinema conquering the landscapes of nature]. In P. de Theux (Ed.), *Médias plus verts que nature L'exploitation du thème de l'environnement dans les médias*, novembre (pp: 35–41). Brussels: Media Animation.

Books & ideas. (n.d.). *Towards an ecological cinema*. Retrieved from: https://booksandideas.net/Towards-an-Ecological-Cinema

Boyle, D. (Director). (2010). *127 Hours* [Moving Picture]. Fox Searchlight Pictures.

Carnahan, J. (Director). (2011). *The Grey* [Moving Picture]. Open Road Films.

Carter, S. G. (2008). Constructing an independent Moroccan nation and national identity through cinema and institutions. *The Journal of North African Studies, 13*(4), 531– 559.

Carter, S. G. (2009). *What Moroccan Cinema? A Historical and Critical Study, 1956–2006*. Lexington Books.

Chahal Sabbagh, R. (Director). (2003). *Le cerf-volant* [The Kite]. [Moving Picture]. Lebanon : Ognon Pictures, Ulysse Films, Arte France Cinéma (France), Leil Production (Liban).

Cinema Punk. (2022). A quick guide to visual style. Retrieved from: https://cinepunked.com/2022/09/23/a-quick-guide-to-visual-style/.

Coen, J., & Coen, E. (Directors). (2007). *No Country for Old Men* [Moving Picture]. Miramax Films.

Coppola, F.F. (Director). (1979). *Apocalypse Now* [Moving Picture]. United Artists.

Costa-Gavras (Director). (1969). *Z*. [Moving Picture]. France/Algeria.

Curtiz, M. (Director). (1942). *Casablanca* [Moving Picture]. Warner Bros.

Dinur, E. (2017). The Filmmaker's Guide to Visual Effects the Art and Techniques of VFX for Directors, Producers, Editors, and Cinematographers. Routledge

Dorlin, O. (2019). *Toward an Ecological Cinema. Public Books*. Retrieved from: https://www.publicbooks.org/toward-an-ecological-cinema/

Earth.Org. (n.d.). *The 21 Best Environmental Films of 2023*. Retrieved from: https://earth.org/best-environmental-films/

Euronews (2021, October 28). *The evolution of Arab Cinema*. Retrieved from: https://www.euronews.com/culture/2021/10/28/the-evolution-of-arab-cinema
Filmmaker Freedom. (2017). Visual planning: A guide to the creative process for filmmakers. Filmmaker Freedom.
Filmmaking Lifestyle. (n.d.). *11 Best Moroccan Movies: A Journey Through the Cinema of Morocco*. Retrieved from: https://filmlifestyle.com/best-moroccan-movies/
Filmscalpel. (n.d.). *Landscape, Nature and Environment in War Films*. Retrieved from: https://www.filmscalpel.com/landscape-nature-and-environment-in-war-films/
Forbs. (2022, Feb 28). *How Film Production Is Becoming More Sustainable and Profitable*. Retrieved from: (forbes.com)
Geal, R. (2021). *Ecological Film Theory and Psychoanalysis: Surviving the Environmental Apocalypse in Cinema*. Routledge. Retrieved from
Glaw, X., Inder, K., Kable, A., & Hazelton, M. (2017). Visual Methodologies in Qualitative Research: Autophotography and Photo Elicitation Applied to Mental Health Research. *International Journal of Qualitative Methods, 16*(1). https://doi.org/10.1177/1609406917748215
Goodvest. (n.d.). *La pollution dans l'industrie du cinéma* [Pollution in the film industry]. Retrieved from: https://blog.goodvest.fr/articles/la-pollution-dans-lindustrie-du-cinema
Guggenheim, D. (Director). (2006). *An Inconvenient Truth* [Moving Picture]. Paramount Classics.
Heckerling, A. (Director). (1995). *Clueless* [Moving Picture]. Paramount Pictures.
Hoad, P. (2020, Sep 2). *BFI study calls on film industry to urgently reduce emissions*. Retrieved from: https://www.theguardian.com/film/2020/sep/02/bfi-study-calls-on-film-industry-to-urgently-reduce-emissions
Hurbis-Cherrier, M. (2013). *Voice and Vision, a Creative Approach to Narrative Film and DV Production*. London: Routledge.
Iddins, A. (2017). Producing Public Intellectuals: Shifting Scales and Social Critique in Moroccan Cinema. *Communication, Culture and Critique, 10*(3), 499–517. https://doi.org/10.1111/cccr.12169
IMDa. (n.d.). *100 Best Winter and Snow Movies (2000–2017)*. Retrieved from: https://www.imdb.com/list/ls021435319/
IMDb. (n.d.). *Winter and Snow Movies: The Ultimate List*. Retrieved from: https://www.imdb.com/list/ls066920305/
IMDc. (n.d.). *Cold weather themed films*. Retrieved from: https://www.imdb.com/list/ls054661888/
Jackson, P. (Director). (2001–2003). *The Lord of the Rings* [Moving Picture]. New Line Cinema.
Jimenez M. P., DeVille N.V., Elliott, E. G., Schiff, J. E., Wilt, G.E., Hart, J. E, & James P. (2021). Associations between Nature Exposure and Health: A Review

of the Evidence. *Int J Environ Res Public Health. 30*(18), (9):4790. doi: 10.3390/ijerph18094790. PMID: 33946197; PMCID: PMC8125471.
Joucka, H. (2015, Aug 08). *Moroccan director combines film, opera to tell story from Algerian independence war.* Retrieved from https://jordantimes.com/news/local/moroccan-director-combines-film-opera-tell-story-algerian-independence-war
Kamal. K. (Director). (2002). فيطرازن [Ghost of Nizar]. [Moving Picture]. Morocco.
Kamal. K. (Director). (2005). *La symphonie marocaine* [The Moroccan Symphony]. [Moving Picture]. Morocco.
Kamal, K. (Director). (2014). *Sotto Voce.* [Moving picture]. Maroc/France: JFK Production.
Kechiche, A. (Director). (2007). *The Secret of the Grain.* [Moving Picture]. France/Tunisia.
Kentis, C. (Director). (2003). *Open Water* [Moving Picture]. Lions Gate Films.
Khan, M. (2001). *An Introduction to Arab Cinema.* London: Informatics.
Laabidi, A. (2022). Openness to Taboo Themes in the Moroccan Cinema: 'La source Des Femmes'. *International Journal of Science and Research (IJSR), 11*(2), 466–468.
Leaman, O. (2001). *Companion Encyclopedia of Middle Eastern and North African Film.* New York: Routledge.
Lean, D. (Director). (1962). *Lawrence of Arabia* [Moving Picture]. Columbia Pictures.
Lee, A. (Director). (2012). *Life of Pi* [Moving Picture].20th Century Fox
Lee, S. (Director). (1989). *Do the Right Thing* [Moving Picture]. Universal Pictures.
Le Matin (2014, February 15). "Sotto Voce" de Kamal Kamal rehausse le ton du cinéma marocaine ["Sotto Voce" by Kamal Kamal raises the tone of Moroccan cinema]. Retrieved from: https://lematin.ma/journal/2014/15e-festival-national-du-film-a-tanger_-sotto-voce--de-kamal-kamal-rehausse--le-ton-du-cinema-marocain/196839.html
Lim, J. A. J. (2012, August 7). *Story catalyst: The most exciting scene in the movie?* Retrieved from: Story catalyst: the most exciting scene in the movie? - The Story Department
Loayza, B. (2021, May 27). *'The Mercy of the Jungle' Review: Soldiers Lost Among the Enemy. The New York Times.* Retrieved from: https://www.nytimes.com/2021/05/27/movies/mercy-of-the-jungle-review.html
Lucas, G. (Director). (1977). *Star Wars* [Moving Picture].20th Century Fox.
Macdonald, K. (Director). (2003). *Touching the Void* [Moving Picture].FilmFour .
Malick, T. (Director). (1998). *The Thin Red Line* [Moving Picture].20th Century Fox.
Martin, F., Higbee, W., & Bahmad, J. (2020). *Moroccan Cinema Uncut: Decentred Voices, Transnational Perspectives.* Retrieved from: https://www.researchgate.net/publication/344167699_Moroccan_Cinema_Uncut_Decentred_Voices_Transnational_Perspectives
Marshall, F. (Director). (1993). *Alive* [Moving Picture]. Touchstone Pictures.

Marshall, F. (Director). (2006). *Eight Below* [Moving Picture]. Walt Disney Pictures.

McKee, R. (1997). *Story: substance, structure, style and the principles of screenwriting*. New York, Regan Books.

Mental Health Organization (n.d.). Nature: How connecting with nature benefits our mental health. Retrieved from: https://www.mentalhealth.org.uk/our-work/research/nature-how-connecting-nature-benefits-our-mental-health

Milestone, L. (Director). (1930). *All Quiet on the Western Front* [Moving Picture]. Universal Pictures.

Miller, C. R., & Kelly, A. R. (Eds.). (2017). *Emerging Genres in New Media Environments*. Springer International Publishing.

Næss, P. (Director). (2012). *Into the White* [Moving Picture]. Zentropa International Norway.

Nejade, R. M., Grace, D., & Bowman, L. R. (2022). What is the impact of nature on human health? *A scoping review of the literature. J Glob Health*. doi: 10.7189/jogh.12.04099. PMID: 36520498; PMCID: PMC9754067.

Orlowski, J. (Director). (2012). *Chasing Ice* [Film]. National Geographic.

Oussama, M. (Director). (2002). قودنص ايندلا [The Box of Life]. [Moving Picture]. Egypt.

Oxford Bibliographies. (n.d.). *Historical Film - Cinema and Media Studies*. Retrieved from: https://www.oxfordbibliographies.com/abstract/document/obo-9780199791286/obo-9780199791286-0239.xml

Petersen, W. (Director). (1981). *Das Boot* [Moving Picture]. Columbia Pictures.

Pontecorvo, G. (Director). (1966). *The battle of Algiers*. [Moving Picture]. Italy/Algeria.

Psihoyos, L. (Director). (2009). *The Cove* [Moving Picture]. Lionsgate.

Reid, D R., & Sanders, B. (2021). *Documentary Making for Digital Humanists*. Cambridge, UK: Open Book Publishers. https://doi.org/10.11647/OBP.0255

Renaud, C., &Balda, K. (Directors). (2012). *The Lorax* [Moving Picture]. Universal Pictures.

Rosa, K. M. (2019). *The contours of an exploding territory: landscape and technology in contemporary war films*. (Unpublished PhD Thesis). Federal University of Santa Catarina, Florianopolis, Brazil.

Sadoul, G. (1996). *The Cinema in the Arab Countries*. Beirut: Arab Film and Television Centre.

Schenk, S., & Long, B. (2021). *The Digital Filmmaking Handbook*. Boston: Cengage Learning.

Serghini, J. (2018). Quand le cinéma africain se fait universel. Le cas du réalisateur marocain Kamal Kamal. In U. Fendler& C. Vatter (Eds.), *Cultural Dynamics in African Cinemas of the 21st Century Agents, Formats, Genres* (pp : 53–62). Saarbrücken: Universitätsverlag des Saarlandes, Saarland University Press.

Shafik, V. (1988). *Arab Cinema: History and Cultural Identity*. Cairo: American University Press.

Schirmer Encyclopedia of Film (n.d.). *Arab Cinema.* Retrieved from: https://www.encyclopedia.com/arts/encyclopedias-almanacs-transcripts-and-maps/arab-cinema
Spielberg, S. (Director). (2011). *War Horse* [Moving Picture]. Touchstone Pictures.
StudioBinder. (2019, November 21). *15 Best Inciting Incident Examples in Film for Screenwriters and Directors.* Retrieved from https://www.studiobinder.com/blog/inciting-incident-examples/
StudioBinder. (2020). *Movie Themes: Examples of Common Themes for Screenwriters.* Retrieved from: https://www.studiobinder.com/blog/movie-themes-examples/
Synchronicity Earth. (2020). How can environmental films make an impact? Synchronicity Earth. Retreived from: https://www.synchronicityearth.org/how-can-environmental-films-make-an-impact/
Tbato. H. (2004). *National Cinema in Morocco: questions of establishment and the artistic conscious.* Morocco.
Tlatli, M. (Director). (1994). صمت القصور [The Silences of Palaces]. [Moving Picture]. Tunisia.
The Writing Center. (n.d.). *Film Analysis – The Writing Center.* University of North Carolina at Chapel Hill. Retrieved from: https://writingcenter.unc.edu/tips-and-tools/film-analysis/
Town & Country. (2022). *20 Must-Watch Survival Movies - Top Survival Movies of All Time.* Retrieved from: https://www.townandcountrymag.com/leisure/arts-and-culture/g32189918/best-survival-movies/
Trottier, D. (2014). *Screenwriter's Bible: a complete guide to writing, formatting, and selling your script.* Los Angeles: Silman-James Press.
Understanding Media and Culture. (n.d.). *8.3 Movies and Culture – Understanding Media and Culture.* Retrieved from: https://open.lib.umn.edu/mediaandculture/chapter/8-3-movies-and-culture/
University of Minnesota (n.d). How Does Nature Impact Our Wellbeing? Retrieved from: https://www.takingcharge.csh.umn.edu/how-does-nature-impact-our-wellbeing
University of North Carolina at Chapel Hill (n.d). *Film Analysis.* Retrieved from: https://writingcenter.unc.edu/tips-and-tools/film-analysis/
Vallet, F., Bazille, F., Gadelorge, C., &Laurentin, R. (2022, Septembre 02). *Culture: le cinéma modifie ses habitudes pour préserver l'environnement* [Culture : the cinema changes its habits to preserve the environment]. Retrieved from: https://www.francetvinfo.fr/meteo/climat/culture-le-cinema-modifie-ses-habitudes-pour-preserver-l-environnement_5339632.html

10

Visual Environmental Analysis of Displaying Images of Hunted Animals in Facebook

Fouad Akki, Mohammed Larouz, Hicham Laabidi, and Mehdi El Madani

In addition to the undeniable role that language plays in describing and alleviating environmental issues, ecosystems can also be represented in visuals through mainstream media (Hansen, 2017). Nevertheless, it is only in the recent century that environmental visual communication, as a promising research area, has started to attract the attention of researchers (Hansen & Machin, 2008, 2013). Since Visual Environmental Communication (VEC) examines how visual content communicates and conveys messages about the environment, its principal interest is to understand the contribution that visuals make to the wider social, political, and cultural construction and understanding of the environment (Hansen, 2017).

In this regard, research on the visual representation of the environment has largely focused on media forms such as television news (e.g., DiFrancesco &Young, 2011; Hansen & Machin, 2008; Lester & Cottle, 2009;),environmental advertising(e.g., Ahern et al., 2013; Howlett & Raglon, 1992; Kroma & Flora, 2003; Linder, 2006), as well as newspapers and magazine covers (e.g., Meisner & Takahashi, 2013; Mliless & Larouz, 2020).

Despite the significance of the above studies, social media such as Facebook, Twitter, and YouTube can also be relevant objects for environmental research pertaining to climate change, wildlife management, and green energy; these media forms provide a great chance for users to discursively frame nature through groups, hashtags, pictures, videos, and so forth (Mörner & Olausson, 2017). Unfortunately, studies on environmental visual representation in social media are scarce and scant. Accordingly, the current study aims to contribute to our understanding of how nature is framed by people in their daily lives by analyzing the images of hunted animals in Facebook pages in different sites all over the world.

Review of Literature

Reviewing the literature reveals that nature has always been depicted as a victim of injury that is generated by human crimes and animal abuse. Going back to ancient history, people used to hunt animals for food and ritual purposes. Ranta et al. (2020), for instance, examined Scandinavian petroglyphs (rock paintings) and concluded that they display phases of hunting, killing acts, and confrontational traits.

With the advent of new technologies, hunters worldwide have sought to display their hunting activities through social media to reach communities of hunters and non-hunters. According to Ponjiger et al. (2022), Facebook pages are relied on by National Hunters' Associations in Europe to promote hunting, protect wildlife, and shape public opinion. However, it is noted in other studies that these pages might not always be used to achieve these purposes.

For instance, Kalof and Fitzgerald (2003) explored the representation of dead animals in hunting magazines and found extreme objectification and marginalization of animal bodies, and a number of trophy exhibits hid the animal body behind or beneath weapons and other hunting equipment. Similarly, Maroun (2015) examined the issue of displaying hunted animals on Facebook in the Lebanese context and concluded that trophy hunting and showing off achievements are widespread among hunters along with supporting comments from the public opinion. According to Pospisil (2017), trophy hunting denotes a kind of recreational hunting of nonhuman animals, which is practiced to get their body parts as an exhibition of success or memorial. In this case, animals are not only hunted as food to satisfy peoples' nutrition needs but as a boastful act to display one's abilities and accomplishments as well. Trophy hunting, therefore, has been debated by both the public and experts, and there has been a call to avoid and face its negative effects such as endangering wildlife populations and shrinking native biodiversity (Ghasemi, 2021).

In the same vein, Mkono (2019) investigated social media narratives on Facebook pages to elicit the attitudes toward trophy hunting among African people. The study demonstrated that trophy hunting was seen as a political arena through which the natural and wildlife resources of Africa are exploited and stolen by rich westerners along with the involvement and collaboration of corrupted and greedy African leaders.

Along with the same moral debates on trophy hunting, Batavia et al. (2019) claims that trophy hunting is an unforgivable act and animals, like human beings, have got intelligence and emotions. In addition, they argue, in trophy hunting, the animals are debased, commoditized and "relegated to the sphere of mere things when they are turned into souvenirs, oddities, and collectibles" (p. 3). Additionally, Mörner & Olausson (2017) examined how wild-boar hunting is depicted on YouTube by adopting a qualitative frame analysis. In their study, hunting was depicted as battle, as consumption, and as privilege which suggests a more hegemonic framing of nature that encourages suppression and dominance of humans over nature and non-human creatures.

Following the same line of reasoning, Jepson (2008) delved into the linguistic analysis pertaining to the killing of nonhuman animals and found that it carries with it connotations of brutality and the human right to take the animals' lives. Additionally, McGuigan and Clark-Parsons (2019) examined hunting scenes in the US television and concluded that it is oriented around product marketing, consumerism, and "killing on camera" at the expense of educating the viewers about nature conservation.

In the Polish context, Krokowska-Paluszak et al. (2020) found that hunters have faced a low level of social acceptance in recent years; while people who are familiar with hunting and hunters have a close relationship with them and therefore eat "game meat" support hunting, other non-hunters have negative attitudes towards hunting.

Theoretical Framework

Under the umbrella term of qualitative research and by adhering to framing theory (Goffman, 1974), the current study is meant to analyze how hunted animals are depicted on different Facebook sites all over the world. Within the field of visual environmental communication, frame analysis has already been relied upon in investigating news media (e.g., Mörner & Olausson, 2017; Shehata & Hopmann, 2012). In this regard, investigating Facebook images entails a lot of visual analysis that constitutes excellent material for studying frames of any kind. Such process consists of attaching meaning to events, processes, and phenomena (Goffman, 1974); and visuals arguably serve as powerful cognitive windows (Pan & Kosicki, 1993) through which the world receives meaning (e.g., Olsen, Finnegan, & Hope, 2008). Frames therefore are organized around "a central organizing idea or story line that provides meaning to an unfolding strip of events" (Gamson & Modigliani, 1987, p. 53).

Nevertheless, the field of visual studies, according to Hansen and Machin (2013), is a multidisciplinary domain that draws from a range of disciplines, including linguistics, rhetoric, semiotics, sociology, social psychology, and historical studies. This entails a multimodal model of analysis of semiotic, discursive, and rhetorical attributes of visuals in relation to three major contexts (communicative, cultural and historical) and three main sites (production, content, and consumption) that could be also referred to as sources/production of communication, communication content and audiences, consumption, and social implications of communication.

This perspective combines core elements of semiotics (e.g., Barthes', 1977), notion of denotation, connotation and myth/ideology), with key propositions of critical discourse analysis (e.g., examining how social identities, relations and systems of knowledge are discursively, visually and ideologically structured). Critical discourse analysis takes the perspective of "those who suffer most from dominance and inequality" (van Dijk, 1993, p. 252). Hence, non-human creatures/oppressed are likely to suffer from exploitation and inequality of human beings/ oppressors. Hunted animals in this sense are vulnerable to brutality of humans, and because animals are voiceless and cannot defend themselves, it is ethical to examine such issues.

In accordance with these theoretical frameworks, this study is meant to analyze the visuals of hunted animals on Facebook, along with a qualitative analysis of the comments on hunted animals on Facebook, as a ubiquitous and widespread social media source of communication, and to determine the mast salient frames and themes in the data.

Materials and Methods

The selected materials consist of images of hunted animals on different Facebook sites all over the world. The researchers check these sites and gain access to their visual content. The final sample consists of 200 images that varied in terms of the country of origin. Description of the images and Facebook pages are represented in Table 10.1.

Table 10.1 Description of the Images and Facebook Pages

Images Number	**Country**	**Facebook Page**
40	Morocco	https://www.facebook.com/groups/472445577298692/ https://www.facebook.com/groups/1624460184551538/ https://www.facebook.com/groups/287233288550883/ https://www.facebook.com/WildhuntiM
40	Algeria	https://www.facebook.com/groups/1295890057283878/ Https://www.facebook.com/groups/677079476631985/
40	Lebanon	https://www.facebook.com/lebanesehuntingclub
40	USA	https://www.facebook.com/wehuntusa Https://www.facebook.com/SpeyCreekTrophyHunting/about
40	England	Https://www.facebook.com/profile.php?id=100046173145926

Afterwards, the selected images were analyzed qualitatively to determine the most salient aspects and frames in the data. Along with this, the comments of Facebook users were also included in the analysis. Because Facebook has been a ubiquitous social media platform and the data were already available in the public sphere, no consent was obtained from the groups' members. However, no identifying information was collected or included in the presentation of the findings, and the images that include people's faces were not incorporated into the content of this paper.

Results and Discussion

This section is devoted to the representation of the study's findings; it is organized around the main frames that emerged from the results, where images are depicted as a) commodity and consumption, b) confrontation and battle, and c) privilege and achievement.

Hunting as Consumption and Commodity

Of great significance to hunters is the hunting of animals such as deer, rabbits, hedgehogs, and birds to satisfy people' nutrition needs and to offer more joy to hunting experiences. "Game meat" is perceived as an outstanding and excellent achievement among the hunter's community. It is displayed in the Facebook images either as fresh or cooked.

Most of the comments mention how tasty and gorgeous such food is. It is the most delicious food that "one's heart cannot stand" according to one of the Facebook users. "Game meat" for hunters is a "bonus," which is a sum of money added to hunters' income as a reward for good performance. Happiness of being outdoors carries with it enjoyment and good quality family time. It is also an "assurance of knowing where the food comes from." These comments are likely to encourage the consumption of "game meat" and make of nature an arena of consumption or market where hunters look for biological and rare food; and what was valued requiring protection has become instrumental to provide consumers with utility (Spash & Aslaksen, 2015). This Finding also matches McGuigan and Clark-Parsons (2019) who claim that hunting is oriented around product marketing and consumerism.

Additionally, the visual analysis of the Facebook images reveals that hunted animals are depicted as objects that could be bought for decorative purposes. For instance, pigs' and deers' heads are mummified and put for sale.

Facebook then is viewed as a marketplace for hunters and non-hunters to bargain about the prices of animals' body parts. In this regard, Stibbe (2003) analyzed the discourse of animal industry in which pigs were represented as machines and objects and concluded that this might encourage intensive farming that is inhuman and environmentally damaging. Kalof and Fitzgerald (2003) also explored the representation of dead animals in hunting magazines and found extreme objectification and marginalization of animal bodies. Similarly, Batavia et al. (2019) found that hunted animals are debased, commoditized and relegated to the sphere of mere things when they are turned into souvenirs, oddities, and collectibles. According to Pospisil (2017), trophy hunting denotes a kind of recreational hobby, which is practiced getting the animals' body parts as an exhibition of success or memorial.

Along with these findings, Facebook represents hunted animals along with prey animals such as dogs for the sake of selling them. Dogs and puppies, for instance, are put up for sale. The comments revolve around asking about their prices as well as detailed information about their age, breed, and the seller's location.

Animals such as dogs and birds of prey, including eagles, hawks, vultures, ospreys, buzzards, and falcons, are a source of income for many hunters that put them for sale. These non-human creatures are also trained and used to harvest other animals, which suggests a war of animals against other animals. For the sake of satisfying the hunters' desires and enjoying watching scenes of confrontation and battle between animals, prey animals are therefore utilized to kill their counterparts without any concern about the environment conservation and animals' rights.

Hunting as Battle and Confrontation

Great battles and confrontations as well as brutality scenes are sources of enjoyment for watchers/hunters. One of the comments urges the photographer to "shoot them all" which suggest a great deal of brutality on the part of the hunters. Nevertheless, animals are not the only means that hunters use to harvest their trophies. For instance, the images of hunted animals are posted along with the weapons.

Along with the trophies, images display hunting equipment such as rifles, knives, necklaces, and traps. It seems that hunting is a war against animals in which various materials and all kinds of weapons are used to oppress nature. This frame is attached to the previous one (commodity and consumption) in the sense that weapons are also put up for sale and their prices are negotiated among Facebook users. Discussions revolve around the brand, quality and price of these tools that are sold and bought to torture animals. In this regard, Kalof and Fitzgerald (2003) asserted that a number of trophy exhibits hid the animal body behind or beneath weapons and other hunting equipment.

Hunting as Privilege and Achievement

The Facebook visual content shows hunted animals as trophies that hunters hold to show off achievement and privilege. Going beyond the frames of consumption and making of nature a source of money, hunted animals are also conceived as rewards. In most cases, for instance, pigs are not hunted to be cooked especially in Islamic contexts (the cases of Morocco and Algeria) in which the pig's meat is forbidden. Hunting pigs, for instance, in these contexts is likely to generate the joy of showing off achievement through these torture and killing acts.

To show great achievement, the size of the trophy is emphasized. Comments give much value to big trophies by displaying their large bodies. In American settings, for instance, big trophies signify great achievement; the size of deer's antlers is highlighted in the images to show the large size of the trophy.

To show the importance of size in trophies, hunted animals are displayed in order of size, from the biggest to the smallest ones. As the size increases, the importance of the trophy increases too. Big trophies for hunters therefore is a good income and a source of "game meat."

Most of the comments on these trophies give much importance to the trophies' size. For instance, Facebook users think that one of the trophies is "the first or second largest in the state" and a "real monster," along with encouraging and rewarding comments such as "Funny story," "Great," "Wow," "That is some rack!," "Congratulations," and "Great goodness." Other comments emphasize how scant and rare are the hunted animals; these include "it will make a great mount," "I've seen two bucks in my lifetime," "very rare," and "Wow that's something you don't see every day, good job." Hence, endangered and rare animals are preferred among the hunters' community; they are viewed as a symbol of courage and bravery. In addition to the notion of size, the number of trophies is also given much value among the hunting community.

In the Lebanese context, for instance, hunters do encourage each other to harvest a huge number of trophies. As the above figure shows, harvesting a large number of endangered birds is widespread among the Lebanese hunters; and privilege and achievement are attached to the number of the trophies. In this regard, Maroun (2015) examined the issue of hunting animals in the Lebanese context and claims that many species are under the threat of extinction due to poaching, which is the illegal and unregulated trafficking and killing of the wildlife. Along with supporting comments from the public opinion, mainly from the hunting community, and for the sake of showing off achievements, endangered species are still victims, without strict measures that could blame those involved in these criminal acts. In this regard, Mliless and Larouz (2020) examined print media coverage of environmental issues in the Covid-19 pandemic era in the Moroccan context and asserted that the poachers convicted of illegal hunting during that period were released; and media has not sufficiently covered these issues in the sense that the violators' identity was erased in the newspaper articles. Similarly, in displaying the images of hunted animals in the current study, it is also observed that some hunters tend to hide their faces.

The hidden faces might suggest the idea that something must be wrong with these killing acts; people hide their faces to avoid being responsible. Therefore, this gives the impression that the doer is unknown. Such findings are in agreement with studies that highlight the deletion of the agent in talking about environmental issues (e.g., Farooq & Umar, 2021; Kahn, 1992; Mliless & Larouz, 2018; Stibbe, 2015).

Using non-visual linguistic patterns such as passive voice and nominalization, the doer of the action is erased and the deed is given central importance (Kahn, 1992). It is rather analogous to the use of euphemistic language by politicians and government officials, who state facts in a soulless, passive voice, therefore not taking any moral responsibility for their actions (Farooq & Umar, 2021). In the visual content, erasure of the agent/ doer is maintained by deleting the hunters' faces in the images and highlighting the trophies as well as prey animals along with the weapons and other tools that are used to kill animals.

Finally, the outdoors and tradition, according to hunters, carry with them memories, enjoyment, and getting out in the wild with friends and family. Additionally, the passion of the outdoors and having the honor to pass it to the hunters' kids as their parents did with them generates feelings of joy and pride. Instead of making children aware of the issues of environment degradation, the hunters demonstrate readiness to teach their children this hobby and pass it to the next generation, which reassures the continuity of endangering rare species. According to most of hunters and hunting community, hunting is a lovely experience that involves preparation, training, pursuit, patience, and rewards.

Conclusion

Due to the scant nature of environmental visual research, the current study examines the displaying of images of hunted animals in the Moroccan, Algerian, Lebanese, American, and English settings by analyzing the visual content of ten Facebook pages. The findings reveal that hunted animals are perceived as consumption and commodity, confrontation and battle, and as privilege and achievement. These findings present nature as oppressed, dominated, and as something to be consumed and enjoyed by human beings, which is also found in similar studies (Batavia et al., 2019; Mörner & Olausson, 2017; Pospisil, 2017)

In addition, nature is conceived as a marketplace to generate income for hunters by consuming "game meat" and selling animals' body parts, prey animals, and weapons and other equipment that are used to endanger the life of many species. Additionally, confrontation scenes generate the joy of watching real battles and brutal scenes; and the size and number of the trophies are signs of great hunting performance. In this regard, hunting endangered animals signifies great achievement, and what should be preserved and protected is preferred among the hunting communities. Accordingly, because these could be viewed as criminals' acts, hunters sometimes delete their faces while displaying the images of hunted animals.

These findings give rise to the requirement of lessening the negative impact of unethical hunting and poaching to reassure the protection of endangered species in the near future. With the emergence and advent of social media, such as Facebook, for instance, more anti-hunting pages are to be created to disseminate awareness among the public. Animals, as the images display, are oppressed and helpless creatures that really need protection. In this regard, Animal Rights Associations should make alarms to protect wildlife and regulate the domain of hunting.

While the findings are compelling here, this study suffers from a set of limitations. Given the large scope of visual content, videos, for instance, were not incorporated in this visual analysis. In this regard, videos can provide a rich source of visual data that include scenes of confrontation, either between humans and non-humans or between prey animals and the trophies, in the wildlife. Due to the complex and difficult nature of analyzing visual content, other sophisticated methodological approaches are also needed to examine the issue at hand. For instance, quantitative research designs can be adhered to delve into this interesting research area.

References

Ahern, L., Bortree, D. S. and Smith, A. N. (2013). Key trends in environmental advertising across 30 years in National Geographic magazine. *Public Understanding of Science, 22*(4),479–494. https://doi.org/10.1177/0963662512444848

Barthes, R. (1977). Rhetoric of the image in S. Heath (ed.), *Image, music, text: essays selected and translated by Stephen Heath*. Fontana, pp. 32–51.

Batavia, C., Nelson, M. P., Darimont, C. T., Paquet, P. C., Ripple, W. J., & Wallach, A. D. (2019). The elephant (head) in the room: A critical look at trophy hunting. *Conservation Letters, 12*(1). https://doi.org/10.1111/conl.12565

DiFrancesco, D. A. and Young, N. (2011). Seeing climate change: The visual construction of global warming in Canadian national print media. *Cultural Geographies, 18*(4), 517–536. https://doi.org/10.1177/1474474010382072

van Dijk, T. (1993). Principles of critical discourse analysis. *Discourse and Society, 4*(2), 249–283.

Farooq, W., & Umar, H. (2021). The deletion of the human agent in environmental science discourse: An ecolinguistic study. *University Of Chitral Journal of Linguistics & Literature, 5*(2), 89–111. https://doi.org/10.33195/jll.v5iII.300

Gamson, W. A., & Modigliani, A. (1987). The changing culture of affirmative action, pp. 137–177 in Braungart, Richard D. (ed.) *Research in political sociology: A research annual.* JAI Press.

Ghasemi, B. (2021). Trophy hunting and conservation: Do the major ethical theories converge in opposition to trophy hunting? *People and Nature, 3*(1), 77–87. https://doi.org/10.1002/pan3.10160

Goffman, E. (1974). Frame analysis: An essay on the organization of experience. Harper & Row.

Hansen, A. (2017). Using visual images to show environmental problems. In *The Routledge handbook of ecolinguistics* (pp. 179–195). Routledge.

Hansen, A. and Machin, D. (2008). Visually branding the environment: Climate change as a marketing opportunity. *Discourse Studies, 10*(6): 777–794. https://doi.org/10.1177/1461445608098200

Hansen, A. and Machin, D. (2013). Researching visual environmental communication. *Environmental Communication: A Journal of Nature and Culture, 7*(2), 151–168. https://doi.org/10.1080/ 17524032.2013.785441

Howlett, M. and Raglon, R. (1992). Constructing the environmental spectacle: Green advertisements and the greening of the corporate image.*Environmental History Review, 16*(4), 53–68.

Jepson, J. (2008). A linguistic analysis of discourse on the killing of nonhuman animals. *Society & Animals, 16*(2), 127–148. https://doi.org/10.1163/156853008X291426

Kahn, M. (1992). The passive voice of science: Language abuse in the wildlife profession. *The Trumpeter: Journal of Ecosophy, 9*(4), 152–154.

Kalof, L., & Fitzgerald, A. (2003). Reading the trophy: Exploring the display of dead animals in hunting magazines. *Visual Studies, 18*(2), 112–122.

Krokowska-Paluszak, M., Łukowski, A., Wierzbicka, A., Gruchała, A., Sagan, J., & Skorupski, M. (2020). Attitudes towards hunting in Polish society and the related impacts of hunting experience, socialisation and social networks. *European Journal of Wildlife Research, 66*(5), 73. https://doi.org/10.1007/s10344-020-01410-0

Kroma, M. M. and Flora, C. B. (2003). Greening pesticides: A historical analysis of the social construction of farm chemical advertisements. *Agriculture and Human Values, 20*(1), 21–35.

Lester, L. and Cottle, S. (2009). Visualizing climate change: Television news and ecological citizenship. *International Journal of Communication, 3*, 920–936.

Linder, S. H. (2006). Cashing-in on risk claims: On the for-profit inversion of signifiers for "Global Warming." *Social Semiotics, 16*(1), 103–132.

Maroun, S. (2015). Hunting crimes in Lebanon facing government and public indifference. In *dA Derecho Animal: Forum of Animal Law Studies* (Vol. 6, No. 2, p. 011). https://doi.org/10.5565/rev/da.88

McGuigan, L., & Clark-Parsons, R. (2019). This kill shot is brought to you by…: An analysis of hunting television in the United States. *Critical Studies in Television, 14*(1), 55–73. https://doi.org/10.1177/1749602018816931

Meisner, M. S. and Takahashi, B. (2013). The nature of time: How the covers of the world's most widely read weekly news magazine visualize environmental affairs. *Environmental Communication: A Journal of Nature and Culture, 7*(2), 255–276. https://doi.org/10.1080/17524032.2013.772908

Mkono, M. (2019). Neo-colonialism and greed: Africans' views on trophy hunting in social media. *Journal of Sustainable Tourism, 27*(5), 689–704. https://doi.org/10.1080/09669582.2019.1604719

Mliless, M., & Larouz, M. (2018). An ecolinguistic analysis of environment texts in Moroccan English language teaching textbooks. *International Journal of Research in Environmental Studies, 5*(2018), 103–116.

Mliless, M. & Larouz, M. (2020). Print media coverage of environmental issues in the Covid-19 pandemic: An ecolinguistic analysis. *Journal Arbitrer, 7* (2), pp. 182–202. https://doi.org/10.25077/ar.7.2.182-202.2020

Mörner, C. & Olausson, U. (2017). Hunting the beast on YouTube. The framing of nature in social media. Nordicom Review 38(1), 17–29. https://doi.org/10.1515/nor-2016-0038

Olsen, L. C., Finnegan, C. A., & Hope, D. S. (2008). Visual rhetoric. A reader in communication and American culture. SAGE.

Pan, Z. & Kosicki, G. M. (1993). Framing analysis: An approach to news discourse. *Political Communication, 10*, 55–75.

Ponjiger, I., Matejević, M., Stankov, U., Marković, V., Ristić, Z., Kovačević, M., & Barović, V. (2022). *Facebook season: A survey of current practices of national hunting associations.* Facebook Pages: The case of the EU. Turizam, 26(1), 22–35. https://doi.org/10.5937/turizam26-29669

Pospisil, H. (2017). Trophy hunting. In J. Urbanik & C. L. Johnston (Eds.), *Humans and animals: A geography of coexistence* (pp. 330–332). ABC-CLIO.

Ranta, M., Skoglund, P., Persson, T., & Gjerde, J. M. (2020). Hunting stories in Scandinavian rock art: Aspects of 'tellability' in the north versus the south. *Oxford Journal of Archaeology*, 39(3), 228–246. https://doi.org/10.1111/ojoa.12197

Shehata, A. and Hopmann, D. N. (2012). Framing climate change. *Journalism Studies, 13*(2), 175–192.

Spash C.L, & Aslaksen I. (2015). Re-establishing an ecological discourse in the policy debate over how to value ecosystems and biodiversity. *Journal of Environmental Management, 159*, 245–253.

Stibbe, A. (2003). As charming as a pig: The discursive construction of the relationship between pigs and humans. *Society & Animals, 11*(4), 375392.

Stibbe, A. (2015). *Ecolinguistics: Language, ecology and the stories we live by.* Routledge.

11

The Rain Sound as an Affective Force That Unites Humans and Nature

Jaouad Achtitah

Natural phenomena shape our affective lives, from our most banal moods and background feelings to more complex cognitive processes. The affectiveness of nature takes us from our limited scope of self to the wider context of planet. This could be explained by the rich interdisciplinarity of affect theory that identifies the reciprocal effect between self and the natural world. This theory explains how affect circulates between natural phenomena, bodies, and ideas. It clarifies how that circulation occurs to bring to light the powerfulness of nature in shaping our affective experience. In this sense, cultural theorists, such as Brian Massumi and Kathleen Stewart, go for what Baruch Spinoza has stated about affect as "virtual" and "autonomous," something which has the ability to connect our body and soul to the web of life. Antonio Damasio similarly highlights how natural phenomena matter in shaping our cognition and emotion. This makes of nature an agentive part that is deeply rooted in the self's fabric. A more environmentally oriented theory of affect has been maintained by Nigel Thrift who sheds light on the idea of "atmosphere" as a product of natural and affective forces. Affect then is endowed with features of openness and inextricability, and embodied capacities that arise and circulate as intensities among assemblages, which explains the affectiveness of the human-nature relationship. Accordingly, there is no room left for division between mind, body, and nature when there is an affective connectedness between the mental systems, bodily senses, and natural phenomena. In this sort of connectedness nature is an active and affective agent that acts and whose actions have serious consequences for human thinking, feeling, and being. One should account for the affectiveness of the natural phenomena in more careful terms as far as our mental and biological dimensions are concerned (Bladow & Ladino, 2018, pp. 2-3-5-6-7). In more specific terms, the affectiveness of natural phenomena could stem from sounds that are, unlike the human ones, unknown to our linguistic and reasoning patterns. The natural sounds are deeply connected to our psyche, and they affect the remote corners of the mind. The rain sound, for instance, as a natural phenomenon finds its way to the depths of the mental sections and gets us once again amid nature's affective relationships. As a matter of consequence, the rain sound is seen as a natural event that endows us with an affective experience that takes us back to unity with the web of life.

Affect might provide us with an account of how the rain sound influences us and accentuates our relationship with nature. The rain sound then has affective intensities that reach our body and mind. It is a natural phenomenon that proves its agency by affecting the self's internal processes and external connections. The assumption here, as Bondi (2016) assert, is that the naturalness of the rain sound makes it so affective on our body and soul. That means there is an affective involvement with that natural sound. Another point to highlight is Edward O. Wilson's theory of biophilia that stresses our attachment to the natural world, which could be really so supportive to the affectiveness of the rain sound (Duncan, 2018). It is important to bring to light Theodore Roszak's "ecological unconscious," too, as it could emphasize our pre-personal and unconscious connections with the rain sound (Fisher, 2013, p. 4). Based on this, the study of the rain sound might enlighten our perception of the human-nature relationship, and it might get us to account for the affectiveness of the web of life in which we are enmeshed. The rain sound could be perceived as a natural "symbol of some inner process [or] it is rather the embodiment of the soul that nourishes us all" (Fisher, 2013, p. 9). There is certain affective connectedness between that natural sound and self, which has actually required deeper scrutiny. Indeed, further investigation needs to be conducted about how the acoustic aspects of the rain pass from our bodily senses to affect the depths of the psyche. By doing so, another dimension, which is typically affective, would appear to strengthen the human connection with the natural world. Here, the argument is that the rain sound is an affective force that can unite humans and nature. Two fundamental questions are therefore raised: How does the rain sound affect the human psyche? How does the rain sound affect the human-nature relationship?

The Affective Quality of the Rain Sound

The rain sound affects the mental processes on many levels. This is understood in a larger context that the mind is part and parcel of the planet. Based on Daniel

Siegel's standpoint, the rain sound is an intrinsic aspect of the planet's effect on the mind. The cognitive processes connect to the acoustic quality of rain whose patterns go beyond the linguistic and logical outline of human logic. That is, rain provides another source of sound to the mind, which is unknown to what people usually hear or listen to. From Paul Shepard's perception, exposure to the rain sound is characterized by an embodied participation and sophistication that enshroud the mental processes (Duncan, 2018). Then, the background auditory processing of rain could affect perceptual and cognitive processes including emotions, thinking, and memorizing. The affective influence of the sound may result in beneficial consequences for mental arithmetic and performance and cerebral alertness (Sakuragi, 2019). In other terms, careful listening to the rain gradually makes the boundaries between the mind and sound become far more irrelevant than the subtle relationship between the very sound and the mental processes. There is now a growing appreciation and evidence of the positive mental quality of experiencing the rain sound. A good deal of meta-analysis has reported the affectiveness of listening to rain, emphasizing mental improvement and recovery (Duncan, 2018). This sort of acoustic, and therefore mental, affectiveness coming from rain can be traced back to Roszak's theory of "ecological unconscious" which insists on the connectedness between the mental processes and natural phenomena including sounds (Fisher, 2013, p. 4). One might state that the rain sound, though it is not linguistically and meaningfully decoded, is connected to some deep corners of the mind. It fills the mind with emotions and perspective. As a matter of consequence, the rain sound can no longer be considered, if we listen carefully, an outer part of the mind.

The affective quality of the rain sound triggers a large part of memory. The depth psychologist Carl G. Jung's archetypal conception of the world might provide a good understanding of how our memory is aroused while listening to rain. It could be stated that the rain sound has deep archetypal, and certainly ecological, roots in the human memory. This implies an inherent sense of affective reciprocity between the rain sound and memory. In this respect, Proverbio et al. (2018) assert that the background auditory stimuli of rain sounds interfere with memory encoding (p. 2). The affectiveness that ties memory to the rain sound could be conceived in John Livingston's conception of "a deep and primal need" for other forms of life (Fisher, 2013, pp. 7–173). Livingston draws our attention to the biological memory that has had a connection with natural sounds including that of rain ever since. Strong emotions, which have to do therefore with our memories, might be elicited in line with the rain sounds.

Moreover, this sort of affective communication between memory and the rain sound could be conceivable in what Robert Romanyshyn has developed as "alchemical hermeneutics." From Romanyshyn's perspective, there is certain affective communication between that natural phenomenon which is the rain sound and that psychic part which is memory, in a way that our conceptual mind cannot clearly follow (Duncan, 2018, p. 38). As a matter of course, this proves the work of affect whose intensity makes of the rain sound a memory trigger. The rain sound could certainly seep into the depths of memory and affect it in different ways. Therefore, the rain sound has got an affective quality that extensively arouses memory.

The unconscious part of the mind could be affected by the rain sound. This is pertinent to the Jungian notion of "the unconscious identity," which leads us to the idea that the rain sound is part of that very identity. Roszak's "ecological unconscious," which is based on Jung's "the collective unconscious," is also relevant here by highlighting our inherent sense of reciprocity with natural phenomena such as the rain sound. Based on this unconscious connection between mind and the rain sound, affectiveness becomes clear. In this context, the unconscious is understood as a relational and interactive phenomenon which the rain sound could affect. The experience of the rain sound definitely reaches the unconscious whose conceptualization is completely different from the Freudian one in this train of thought. The unconscious here is not a container of unwanted psychic contents, which makes it a kind of separate mind within the mind and body; however, the unconscious is seen as an intrinsic element of the bodily as well as affective experience of natural sounds such as that of rain. The affective experience of the rain sound creates a mode of entanglement in which the unconscious "life" becomes intensely somatic. That is, when the acoustic content of rain passes out of focal awareness, it does not remain all neatly formed as it is. It does not get filed away as a detached entity, but it rather melts back into the flow of our bodily, mental, and certainly unconscious experiencing, to become part of the complex and holistic felt reality of our continuous, affective interaction with the world (Fisher, 2013). For that reason, the unconscious is basically connected not only to images as Jungians might do but also to the sounds of the natural phenomena, with the sound of rain in particular, which could be substantially affective.

Certain positive emotions could be provoked by the effectiveness of the rain sound. Based on Richard Louv's perspective, the rain sound is a natural phenomenon that has a deep positive effect on the human emotions. He introduces a good deal of scientific and speculative evidence that proves the positive affect resulting from natural sounds such as that of rain. The rain sound is experienced as a safe, natural stimulus which enhances a kind of positivity in how we perceive the world. In other words, the rain sound can be considered an emotional, and psychological by extension, counterpart that soothes us (Louv, 2013). According to Proverbio et al. (2018), it is so possible that the rain sound might boost positive emotional arousal in view of its perceptual similarity to white noise. More than that, the rain sound might affect arousal and mood positively due to its being a natural sound. These aforementioned scholars emphatically opt for the idea that natural sounds such as that of rain have positive affective influence on cognitive processes by inducing pleasant emotional states. The same rationale has been adopted by Roger S. Ulrich and colleagues when they have affirmed that the rain sound, given its naturalness, induces wakeful relaxation and sustained attention. Due to the cerebral alertness, it enhances, the rain sound could consequently improve our mental and emotional performance with appropriate sympathetic activation. As a matter of course, the rain sound with a monotonous rhythm and moderate intensity may have positive affective energy that would increase caution level and calmness (Sakuragi, 2019). Thus, one might state that the rain sound satisfies our deep emotional need for connectedness and warmth in nature. The rain sound awakens in us that biophilic principle that we are evolutionarily predisposed to be more emotionally contented when we listen to natural sounds (Duncan, 2018). In short, the positive emotional states the rain sound provokes in us are evidence of our deep need to listen to more-than-human noise.

The affectiveness of the rain sound could be a healing energy that facilitates recovery from psychological stressors. The attention restoration theory developed by psychologists Stephen and Rachel Kaplan is based on evidence that natural sounds are very effective for restoring attention; therefore, the rain sound is an affective force and restorative energy for the brain by having that effect of relaxing the prefrontal cortex. In this sense, the rain sound could be categorized with the stimuli that attract "undirected attention" or soft focus which requires no mental effort and is calming, allowing the mind to clear and settle into a less directed, more reflexive and relaxed state. As Ulrich puts forward, stress recovery could be achieved through the activation of a stress-reducing reflex in the limbic system of the brain after exposure to the rain sound. This could be further explained in terms of the connectedness and affectiveness which exist between the limbic system, automatic nervous system, and unconscious memory where preverbal traumatic experiences are "stored," which could be calmed by the deep brain relaxation that is brought in turn by listening to the rain sound (Duncan, 2018). Concomitant with this, exposure to that natural sound essentially causes, as the growing body of research proves, rapid reduction in psycho physiological stress indicators, affecting heartbeat, adrenaline, and tension (Louv, 2012). Then, the rain sound has an affective capacity and healing energy, as Cassandra Gould van Praag and colleagues assert, in reducing stress or pain to provide relaxation instead, which has major positive effects on cognitive processes. Though the mechanism has not been clearly perceived, the rain sound might affect negative mood by alleviating anger and hostility (Sakuragi, 2019). As a result, the affectiveness of the rain sound could be conceived as a healing energy that massages the mind, especially the anxious one.

How the Rain Sound Affects the Human-Nature Relationship

The rain sound has emotional characteristics that always remind us of our connection with nature. These characteristics make us open to the potential of listening to the rain and engaging with the natural world. This might be understood, as Henry Corbin states, as an invitation to open to the sensuous, emotional, and embodied experience of the rain sound that leads to the realization of a relationship with nature beyond the constraints of logic or category. A deep experience of heart connection with nature could be congruent with an emotional experience of the rain sound, in which the Cartesian dualism is totally irrelevant. It seems possible that the rain sound creates a meaningful, emotional context where we have opportunities to listen to a natural voice that can encourage a gradual awareness of the interconnectedness between self and the world (Duncan, 2018). In view of that, experiencing the rain sound could have two basic aspects: the sound in itself and the subsequent emotional state that entails bodily sensitivity and therefore closeness to nature. According to the Buddhist teacher Chogyam Trungpa's words, "human being has a basic nature of goodness . . . [when] we hear a beautiful sound, we are hearing our own basic goodness," the emotionality and beauty of the rain sound awaken our own basic and intrinsic connection with our inner and outer nature (Fisher, 2013). However, if one fails neither to feel nor realize the emotional characteristics of the rain sound, the gap between humans and nature would become wider. For the rain sound could be considered a natural voice, the loss of these very characteristics would have serious ramifications to the human-nature connection. This is clearly what Jung affirms as he argues that "man . . . has lost his emotional "unconscious identity" with natural phenomena . . . No voices now speak to man . . . nor does he speak to them believing they can hear. His contact with nature has gone, and with it has gone the profound emotional energy that this symbolic connection supplied" (Fisher, 2013). The inability to realize the emotionality of the rain sound is in itself so hazardous that one is blind to their closeness to nature. As a result, the connection, or the symbolic one in Jung's words, with nature is full of emotional energy, and the sound of rain is one fundamental aspect of that energy.

Lack of the rain sound very likely affects our emotional states negatively, which proves the firm connection between self and nature. Lack of the rain sound means lack of one fundamental, natural phenomenon which highly affects our emotions and reminds us of our biophilic relation with the world. This brings to light Louv's notion of "nature-deficit disorder" which describes the negative emotional costs of alienation from nature, which could be a consequence of, for instance, our diminished senses that fail to listen to the rain. Another point to stress is that when there is no rainfall which means no rain sound, emotional distress comes up as an outcome which evidently highlights the indispensable role of nature (Louv, 2013). Based on Matthew et al. (2021) argument, emotional states are so possibly to be affected by the rain sound over different time scales. For instance, emotional distress or exacerbations might occur during a period of low rainfall and little rain sound, while severe depression might be marked after years of prolonged drought. Through this affective relationship between the rain sound and emotional states our connection with nature becomes clearer and stronger. In this way, lack of the rain sound entails serious emotional ramifications, for it is considered an active and affective part that draws us closer to nature. The lack of it implies the absence of a thread that connects us to the world, which leads to undeniable states of distress and anxiety. Further, it is quite possible to tackle the archetypal role of the rain sound here, which means that its lack negatively affects the human emotions and relation with the natural world. The lack of that sound, more or less, upsets us and makes us deaf to that acoustic symbol which "embodies the poetic expression of ways of being and relating to others" in nature "where a beautiful and awesome otherness should have been encountered" (Fisher, 2013, p. 8). In other words, as the rain sound is part of nature it is also part of us; otherwise, the lack of it would go unnoticeable. Our sense of belonging to the natural world is affected once we no longer hear the rain and once we no longer feel comfortable about that. In a nutshell, lack of the rain sound falls disproportionately with its negative affective energy on human beings who are themselves phenomena of nature.

The affective quality of the rain sound affirms the human-nature connectedness. Bessel Van der Kolk argues that the brain is primarily "an organ of attachment" and that humans are continually engaged in an affective connectedness with the natural world around us, in which the rain sound is a fundamental component (Duncan, 2018, p. 59). That is, the human brain is situated in a web of affective relations with the natural phenomena, and the rain sound that affects our psyche in a way or another helps us listen, observe, and feel nature around us. It means that the affectiveness of the rain sound can carry us from the micro-scale of individuality to the macro-scale of nature, and it can therefore reawaken our sense of connectedness with the whole. From Damasio's perspective, there is certain emphasis on the idea that the rain sound, given its naturalness, matters in shaping both cognition and emotion, which actually proves the connectedness between the depths of our mind and the natural phenomenon (Bladow & Ladino, 2018). It is this affective quality that makes of the rain sound an alerting natural call connected to the depths of the human psyche. As it has been noted, the association between the rain sound and the human-nature relationship is based on that complex, affective quality which follows a number of pathways, from the natural phenomenon to the bodily senses, to the brain, to limbic system, to the cerebral cortex, to attention, perception, emotion, memory, consciousness, and bodily awareness in turn (Yap et al., 2021). Thus, it is necessary to account for the affective influence of the rain sound on us in more careful terms as we long for more connectedness with the natural world. Therefore, with the affective quality of the rain sound in mind, the human-nature connectedness is inevitable.

Our emotional attachment to the natural world could be highlighted by the rain sound. William James has put forward his theory of emotional attachment to the world around us in the 1884 essay "What Is an Emotion?" in which he suggests that "bodily changes follow directly the perception of the exciting fact, and that our feeling of the same changes as they occur is the emotion;" therefore, the natural, acoustic quality of rain is an existing fact that provokes changes in us, making us conscious of that emotional attachment between self and nature (Bladow & Ladino, 2018, pp. 38–9). Our openness to be emotionally reverberated by the natural sound of rain suggests our predisposed emotional susceptibility, readiness, and attachment to nature. Then, the rain sound makes us aware of our suprapersonal connections, connections that go beyond our limited self to relate us emotionally to the web of life. As Jung (1933) asserts, it is important to perceive of a natural phenomenon such as the rain sound as something quite "universal" and attached to our psychic mechanisms including emotions. The rain sound could be considered an emotional experience, but it is not peculiar to ourselves since that would be wrong, too personal, and tends to exclude us from the natural connectedness. As a matter of course, as it reawakens our biophilic, emotional attachment and belonging to nature, the rain sound affects us in terms of rekindling personal consciousness in the light of suprapersonal consciousness which is based on the sense of connectedness. Consequently, the rain sound makes us listen not only to ourselves but also to the natural phenomena beyond our bodily limitedness, which enlightens and strengthens our emotional attachment to the web of life.

The rain sound, through its affective energy, is capable of connecting mind, body, and nature. First of all, a good deal of biological thinking has been trapped in a Newtonian worldview which is determined by the mechanistic science and linear paradigms that study mind, body, and nature separately, disregarding any sort of affective connectedness that might result, for example, from the rain sound. This is science that stubbornly follows the classical approach of data collection, categorization, and taxonomy, in which consciousness, bodily awareness, and natural phenomena such as the rain sound are believed to be separate entities. However, the rain sound as a natural phenomenon and an affective force is able to prove the systemic connection between mind, body, and the world. From the scope of systems thinking theory, division and linearity have nothing to do with the premise of this paper since the focus here is the affective connection the rain sound arises between self and the web of life. By putting the Cartesian dichotomies aside, the rain sound brings us to the recognition of the connection between the mental systems and the natural systems. Here, the work of affect becomes clearer as it circulates through sounds, bodies, and mind and creates a shared atmosphere. The acoustic aspects of the rain could be complex, affective patterns that are embedded in the systemic matrix of nature that reveals some archetypal relationship with body and mind. Accordingly, the rain sound might be understood as an affective pattern that connects our mind and body to the natural world. This has to do with Gregory Bateson's "*patterns that connect*," among which the rain sound is conceived of as a pattern that connects our whole being to the natural world (Duncan, 2018, p. 84, 87, 91, 93, 97). Hence, the affective energy that characterizes the rain sound highlights the interactivity and sensitivity of nature with which our mental and physical fabric is connected. By being affected by the rain sound we are getting into nature's complex relationships and patterns that can gradually change human thinking, feeling, and being to be more attuned to the web of life. In short, the fact that the rain sound is affective enough to be captured by bodily senses and mental processes strongly affirms that we are connected to nature on many levels.

Conclusion

The rain sound could be understood as an affective force that unites humans and nature. Its affectiveness resides in its ability to influence our cognitive processes, including consciousness, memory, and emotions, which proves our kinship with natural phenomena. In this respect, our connection with the natural world becomes clearer and stronger as the rain sound is capable of making us aware of the complex, affective relationships in the web of life. The rain sound becomes therefore an affective energy that links our mind and body to nature. Its acoustic aspects are affective patterns that connect self and the world. The rain sound provokes that biophilic connection that unites us biologically as well as emotionally to the natural world. It could be an archetypal element of our unconscious mind and whole being that shall not be apart from the rest of creation. In this sense, the affective quality of the rain sound and its ability to unite humans and nature could be explained in the terms of affect theory. A good deal of evidence supports the affective connection between that natural sound, body, and mind. This is actually the work of affect that is not static but circulating between sounds, bodies, and mind, to create an atmosphere in the end full of connections. Then, the rain sound is a natural phenomenon which is not passive or irrelevant, but it is rather active and affective especially in terms of accentuating human-nature unity. With this in mind, the premise here is that the rain sound might be an affective force that accentuates our deep bond with the natural world. This can be in turn one hint for other studies to be conducted and other natural sounds, or natural phenomena in general, to be considered in terms of affect.

References

Bladow, K., & Ladino, J. (2018). *Affective Ecocriticism: Emotion, embodiment, environment.* University of Nebraska Press.

Bondi, L. (2016). *Emotional geographies.* Routledge.

Duncan, R. (2018). *Nature in mind: Systemic thinking and imagination in ecopsychology and mental health.* Routledge.

Fisher, A. (2013). *Radical ecopsychology: Psychology in the service of life* (2nd Ed.). SUNY Press.

Jung, C. G. (1933). *Modern man in search of a soul.* Harcourt, Brace & World.

Louv, R. (2012). *The nature principle: Reconnecting with life in a virtual age.* Algonquin Books.

Louv, R. (2013). *Last child in the woods: Saving our children from nature-deficit disorder.* Atlantic Books.

Proverbio, A. M., De Benedetto, F., Ferrari, M. V., & Ferrarini, G. (2018). When listening to rain sounds boosts arithmetic ability. *Plos One*, 13(2), e0192296. https://doi.org/10.1371/journal.pone.0192296

Sakuragi, S. (2019, November). Effects of Rain Sound on Mental Arithmetic, Mood, Autonomic Nervous Activity, and Salivary Cortisol. *Neuropsychological Trends, 26*, 63–82. Digital Object Identifier System. Retrieved from: https://doi.org/10.7358/neur-2019-026-saku

Yap, M., Tuson, M., Turlach, B., Boruff, B., & Whyatt, D. (2021, February 1). *Modelling the relationship between rainfall and mental health using different spatial and temporal units.* MDPI. Retrieved from: https://doi.org/10.3390/ijerph18031312

12

The Environment Between Psychology, Identity, and Society

Morocco as a Case of Study

Douae Serraf

The environment is complex and multifaceted, encompassing various aspects of our lives. Choosing to speak about the environment can be challenging for most scholars in psychology, as the majority tend to fall into subjectivity and involve emotions in their approach. The physical environment, also known as nature, connects more with a person's soul and heart than their mind and reason. Gifford (2002), a Canadian psychologist who has made significant contributions to environmental psychology, emphasized that the physical environment is essential in shaping human behavior, attitudes, and well-being (Gifford, 2002).

When we think of nature, we might receive feelings of childhood, picnics, sunny days, and family time. Some may relate the environment to a specific element of nature, such as a tree, wind, rain, air, or an animal. Therefore, the perceived confusion is not confusion but rather a matter of perspective and subjectivity. Kaplan (1983), an American psychologist who conducted extensive research on the relationship between the physical environment and human well-being, described the physical environment as having both positive and negative effects on an individual, depending on their perceptions, expectations, and needs (Kaplan, 1983).

Some individuals may find their environment supportive and fulfilling, while others may experience it as stressful and unpleasant (Kaplan, 1983). One who would perceive nature as optimistic and joyful will eventually meet someone who would not admire it at all, primarily for selecting purposes. Where identities and mindsets differ, perspectives follow and create various complexities. When people have diverse backgrounds, experiences, beliefs, and values, they may perceive things differently and interpret events or situations in their unique way. James G. Greeno, a psychologist and researcher known for his work in the field of cognitive psychology and educational psychology, states in his article "The situativity of Knowing, Learning, and Research: 'The physical environment plays an important role in shaping how people think, feel, and behave'" (Greeno, 1998, p. 5). It is undeniable that nature has a significant role in humans' state of mind and mental health. While the exact mechanisms by which nature influences mental health are not yet fully understood, the growing body of research suggests that nature can play a significant role in promoting human well-being.

Despite being a topic of discussion in numerous academic publications across various countries, the field of environmental psychology remains underdeveloped in Morocco. As Morocco continues its efforts towards achieving sustainability, there is a need for further research across various fields, including the intersection of environment and psychology. The ongoing discussion on this topic in Morocco is far from being concluded. Moreover, in these contexts, inclusive approaches such as environmental psychology and intersectionality are most needed. In this regard, Brown (2010) wrote that in developing countries, the increasing pace of urbanization and industrialization has brought about serious environmental problems, such as pollution, overcrowding, and resource depletion. In these contexts, environmental psychology has a critical role in understanding the relationships between human behavior and the physical environment and developing sustainable and socially just solutions to these problems (Brown, 2010). Therefore, in order to achieve a certain degree of sustainability, an overall understanding of human behavior must be considered. Gifford (2007) states, "The physical environment plays a central role in shaping individual and group identities by influencing how people perceive and experience their surroundings." In other words, Morocco needs similar studies that combine human actions with their environment to advance social awareness and self-esteem.

In this chapter, we argue that adopting environmental psychology in the Moroccan context could result in social understanding and balance in human behavior. Accordingly, this paper attempts to investigate and unveil Moroccans' relationships with the physical environment. It also aims to reveal the degree to which Moroccans are aware of their close connection with nature and its impact on their well-being. This study is conducted in-person, particularly with individuals in Oujda, Casablanca, and Tetouan. This chosen category represents the sample population this paper wants to investigate since they are the representative population concerned with the issue. The chapter is divided into five sections: Section I, an introduction and a review of the literature that gathers several precedent studies regarding the intersection of environmental psychology and identity. Section II, which is the methodology of the study, tackles the following points: the research hypothesis, research questions, approaches, sample of the study, the variables of the study, and the limitations of the study. Section III, the analysis chapter, deals with the analysis of the data collected. Section IV summarizes the study's significant findings, while the last section, Section V, offers the implications and solutions of the study.

Environmental psychology is a field of study that explores the relationship between people and their physical environment. The literature in this field highlights the impact of the physical environment on human behavior, cognition, and emotions, and vice versa. Altman and Werner (1997) describe this relationship as "place attachment," which refers to the emotional bond people form with specific places (Werner, 1997). The physical environment can profoundly affect our thoughts, feelings, and behaviors, according to Aronson et al. (2010). Canter (1977) explains that the psychological sense of place is a crucial aspect of human experience and reflects our relationships with the physical and social environment (Canter, 1977). Gifford (2007) further expands on this concept, stating that environmental psychology provides a framework for understanding the complex interplay between people and their surroundings (Gifford, 2007). Environmental psychology provides a comprehensive understanding of the relationship between people and their physical surroundings, considering the complex intersection between the two. This intersection can majorly impact individuals' well-being, mood, stress levels, and physical health, making it a required field of study. With a focus on understanding the psychological and behavioral effects of the physical environment, environmental psychology can inform design and planning practices to create more livable, sustainable, and inclusive environments.

Nasar (1992) believes that the perception of a city plays a crucial role in shaping the relationship between people and their physical environment, as this can significantly impact how people perceive and engage with those environments. He argues that various social, cultural, and historical factors influence the evaluative image of a city. This suggests that people's experiences with their physical environment are not solely determined by their physical characteristics but are also shaped by the larger cultural and historical context in which they exist. Stokols (1996) focused on translating social ecological theory into design practice, requiring an understanding of the interplay between the physical environment and human behavior (Stokols, 1996). He emphasized the significance of applying social ecological theory to design practices. This requires a comprehensive understanding of how the physical environment interacts with human behavior. The author stressed the importance of integrating this knowledge into design processes to create environments that effectively meet the needs and behaviors of the individuals who will inhabit them.

Some people even say that the physical environment can heal dangerous diseases. Ulrich (1984), a psychologist and environmental researcher, points out the positive impact that views of nature through windows can have on recovery from surgery, reducing stress, and promoting well-being (Ulrich, 1984). He finds that exposure to natural views through windows can benefit recovery from surgery by reducing stress levels and promoting overall well-being. The author concludes that incorporating nature into the design of healthcare environments can positively impact patient outcomes.

"The physical environment profoundly affects the formation of human identity, influencing not just emotions and behaviors but also values, beliefs, and cultural practices. Furthermore, the way that people experience and understand their environment is deeply connected to their sense of identity," as noted by Jackson (1984). He states that "the experience of the environment shapes not just individual self-concept, but also collective cultural identity" (Jackson, 1984). The physical and built environment plays a significant role in shaping human identity, influencing emotions, behavior, and values, beliefs, and cultural practices. Therefore, designers and planners must be mindful of the impact of the physical environment on human identity when creating environments that will impact the lives of those who live and work in them.

In Morocco, the concept of identity encompasses various attributes that distinguish individuals or groups. These characteristics include cultural background, religion, ethnicity, language, and other factors that make up a person's sense of self. Morocco's history as a former colony and its current status as a developing nation also shape the country's national identity. As documented in the High Atlas, "the forest agdal—'patrimony for resilience'—and pastoralists' 'traditional ecological knowledge and adaptive strategies' demonstrate how environmental relations anchor cultural identity and attachment to place. (Auclair et al., 2011; Fernández-Giménez et al., 2021).

The natural environment of the country is also a significant factor in shaping the identities of its inhabitants, particularly in the case of the Berbers, an indigenous group who have lived in the Atlas Mountains for generations and have developed a distinct culture and way of life that is closely tied to their mountain environment. In order to promote cultural understanding and respect for diversity, it is essential to understand how the numerous factors that impact identity interact in Morocco. By doing so, it becomes possible to implement policies and programs that support and respect the unique cultural identities of individuals and groups within the country.

Although literature was inconclusive regarding environmental psychology in Morocco, the floor remains open for further research and exploration to understand this field and Moroccans per se better. The lack of research in the tackled area is a mere sign of progress and future sustainable inquiries. Environmental psychology is a required field of study that has contributed significantly to our understanding of the complex relationship between people and their physical environment. When we learn what triggers our emotions and behaviors, we self-emancipate and work towards sustainability, not only within ourselves, but through our community and among our people. Therefore, the current study aims at setting grounds for inclusive theories and approaches to analyze human behavior in Morocco and further discuss the interconnectedness of humanity's complexities and environment.

Methodology

Research Hypothesis and Objectives

In this chapter, it is hypothesized that the more individuals in Morocco connect to nature and adopt environmental psychology in their livelihood, the more they reach a certain level of self-esteem and social awareness. The study's objective is to investigate the extent to which the physical environment influences Moroccan individuals in the cities of Oujda, Casablanca and Tetouan.

Research Questions

The study attempts to answer the following questions:

1. To what extent do Moroccans feel connected to the natural environment?

2. Can human psychology improve or deteriorate because of the environment in Morocco?
3. To what extent are the Moroccan culture and identity reflected in the built environment?

Research Variables

There are two variables in this study; the dependent variable is Moroccans reaching a certain level of self-esteem and social awareness, and the independent variable is individuals connecting to nature.

Sample Population

In this study, the population sample chosen is Moroccans from different age categories and cities in Morocco. Most of them participated from Oujda, Casablanca and Tetouan. The selection of the cities was grounded in their cultural diversity, as each region serves as a melting pot of different identities. Random sampling is the technique used in order to choose a representative sample. This sampling technique was chosen because it is more objective and allows everyone to participate. The total number of Moroccan citizens exceeds thirty-six million. Therefore, the representative sample, which is equivalent to 15 percent of that total number, is 5,250,000. However, since this study is a mere introduction, the number is far too large. Then, considering the main cities this study would take place, Oujda, Tetouan and Casablanca, the representative sample would be 300.

Sample Description

The current study requires detailed information regarding Moroccans contributing to this research. Therefore, the respondents were asked to specify their sex to check who volunteers most in academic inquiries.

The sample population consisted of 300 respondents, with 52.4 percent being male and 47.6 percent being female (Figure 12.1). Efforts were made to ensure an equal representation of both genders in the sample. Participants came from three different cities in Morocco: Oujda, Casablanca, and Tetouan. 28.6 percent were from Oujda, 28.6 percent were from Tetouan, and 42.9 percent were from Casablanca, showing that Casablanca is a diverse city that attracts people from

Figure 12.1 Respondents' gender. Source: Douae Serraf.

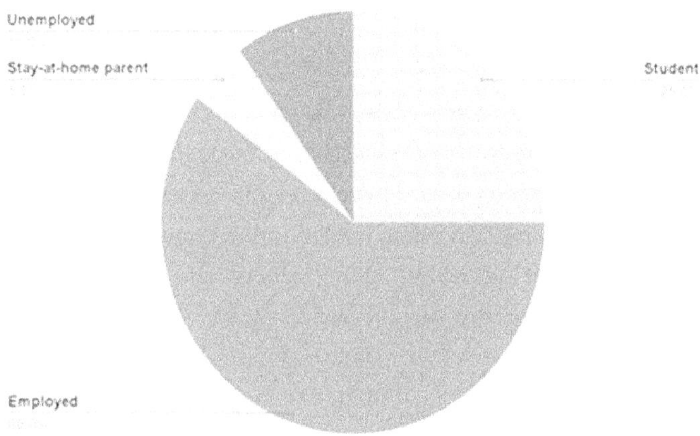

Figure 12.2 Respondents' occupation. Source: Douae Serraf.

across Morocco. The survey also gathered information on participants' ages and occupations. The largest age group was 20–29, accounting for 52.4 percent of the participants, followed by 30–39 with 33.3 percent. 9.5 percent of respondents were forty or older and 4.8 percent were teenagers aged 13–19.

The results of the questionnaire showed that 60 percent of participants were employed, 25 percent were students, 10 percent were not currently employed, and 5 percent were stay-at-home parents. This diversity in occupation provides an opportunity to investigate how the physical environment affects different careers and how an individual's mood and behavior may vary in response to nature (Figure 12.2).

Research Approach

This study will employ both qualitative and quantitative approaches. The quantitative approach will be instrumental in establishing the profiles of Moroccans who are cognizant of environmental psychology, quantifying their connection to nature, and assessing the impact of the physical environment on their well-being. Conversely, the qualitative approach will provide more comprehensive insights into the significance of nature in shaping their identities and its utilization in enhancing their lives. The integration of these approaches will bolster the validity and reliability of the data, thereby enriching the quality of the findings.

Data Collection Procedures

Questionnaire

The questionnaire is the first data collection procedure used in this study. It is a quantitative data collection instrument used to measure the quantity, amount, and frequency of something. One of the strengths of using a questionnaire in this study is to elicit quantitative data related to the frequency Moroccans feel a connection to the natural environment in Morocco, and the extent to which the Moroccan culture and identity are reflected in the built environment.

Interview Guide

Since one of the weaknesses of the questionnaire is its limitation with respect to eliciting in depth answers from the respondents, interviews will also be used to collect data. This qualitative data collection procedure allows respondents the time and scope to talk about their opinions on a particular subject, and it allows the researcher to elicit in-depth information around the topic.

Results

Society in Morocco is highly diverse and complicated. Morocco's position at the crossroads of Africa, Europe, and the Middle East has led to a unique blend of cultures and influences. The natural environment, as well as the built environment, can influence a person's mental health and well-being. The environment in Morocco can have both positive and negative impacts on human psychology and well-being, depending on the specific conditions of the environment and the individual's circumstances. It is essential to consider the impact of the environment on mental health and well-being and take steps to improve environmental conditions if necessary. Understanding the psychological and social factors that influence human behaviors and attitudes towards the environment can provide insight into how to change behaviors and attitudes to promote sustainable practices.

The environment comprises everything around us, including the physical, social, and cultural elements we encounter daily. Our experiences and understanding of the environment can substantially impact our identity and overall well-being. The questions asked are designed to investigate an individual's connection to the environment, specifically their relationship with the natural environment in Morocco, their emotions and actions in response to different weather conditions, how Moroccan culture and identity are reflected in the built environment, and the significance of the environment to their identity and daily behavior. The study's questionnaire consisted of ten main questions on participants' profile, their connection to nature and its influence on their identity, and related topics requiring the respondents to reply by choosing one of three or five options. Some questions required a simple rating of knowledge or awareness of the environment from "limited knowledge" to "advanced," but spaces were provided for them to specify, explain, or state reasons in their own words if they wished to do so.

Moroccans' Perceptions of the Physical Environment

When discussing the relationship between environmental psychology and identity, it is important to include a section on individuals' perceptions of nature. Although the concept may seem abstract, many individuals find it difficult to respond and tend to choose straightforward aspects of nature. Participants were questioned about what immediately comes to mind when they hear the term "physical environment." The responses to this question varied widely, with participants mentioning everything from trees, grass, and flowers to the sky, clouds, sun, and air. Some respondents named specific natural features such as mountains, seas, forests, gardens, or deserts, while others talked about animals like cats, birds, dogs, and fish. Only a small portion of respondents identified the physical environment with the concept of nature or symbolized it with the color green.

Interviewee 16 described the physical environment as having the color green. She said: "*See this question is tricky; the first thing that comes to my mind is the word nature because that's the basic meaning in our society. But I am going to describe it in terms of colors. When you say environment, I see green.*" The interviewee views the word "environment" as being synonymous with nature, which is a common perception in Moroccan society. She also associates the environment with the color green as a representation of what most individuals in Morocco see throughout their entire lives, from textbooks, magazines, to television and social media. Interviewee 6 also responded with the green color representation saying: "*Green is my answer. Trees are green, grass is green, forests are green, parks are green. When you go to mountains, you picture green.*" The color green normally symbolizes nature, growth, and life in individuals' minds, showing their strong relationship with the natural environment and its significance to their understanding of their physical surroundings.

Moroccans' Understanding of the Physical Environment

Respondents were asked to rate their knowledge of the physical environment. They were given six options, from "no knowledge" to "expert." This inquiry aimed to examine the familiarity and understanding of nature among Moroccans. Figure 12.3 offers a close look into the findings.

As demonstrated in Figure 12.3, a large percentage of the respondents acknowledge having limited knowledge of the physical environment. 47.6 percent confess to having only a basic understanding of nature in Morocco,

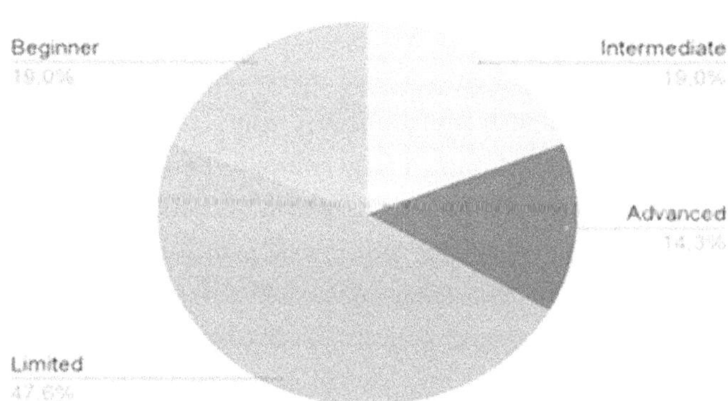

Figure 12.3 Moroccans' understanding of nature. Source: Douae Serraf.

while 19 percent consider themselves to be novices in this area. Another 19 percent believe they have intermediate knowledge. Only 14.3 percent assert to have an extensive knowledge of the physical environment in Morocco. In this context, interviewee 7 states: "*It is hard to say that someone knows nature. No one fully understands nature. We are all in a learning process and most of us abandon nature just because it's vast.*" Another interviewee (18) emphasized it by saying: "*I only know the basic things we see in nature like the trees and flowers and sun and what comes with them. I don't know more than what they taught us and what we learnt in our years.*"

The two interviewees share the sentiment that knowledge of nature is vast and ever-evolving. They both acknowledge that their understanding is limited to what they have learned and experienced, and that there is much more to discover. They seem to believe that many people give up on learning about nature because of its complexity. This highlights the importance of continuing to educate and engage individuals with the natural world to deepen their understanding and appreciation of it.

The findings from Figure 12.4 indicate that a significant number of respondents have limited knowledge about the physical environment in Morocco. Nearly half (47.6 percent) confess to having only a basic understanding, while only a small fraction (14.3 percent) considers themselves to have an extensive understanding. These findings suggest the importance of promoting education and awareness about the physical environment and its role in our lives.

Despite the vastness and complexity of nature, as highlighted by some interviewees, having a basic understanding can serve as a foundation for expanding one's knowledge and appreciation of the natural world.

Moroccans' Connection to Nature

Participants were then asked about the extent to which they feel connected to the physical environment. The relationship between human beings and the natural world has long been a subject of examination. Assessing the level of connection that people in Morocco feel towards their natural environment is crucial in determining the significance and influence nature holds in their lives. This inquiry seeks to uncover the participants' perspectives and encounters with the natural environment in Morocco and to provide insight into their bond with it. The results shown in Figure 12.4 below indicate their responses.

In this section of the survey, participants were presented with five choices to indicate their level of connection to the natural environment in Morocco. The

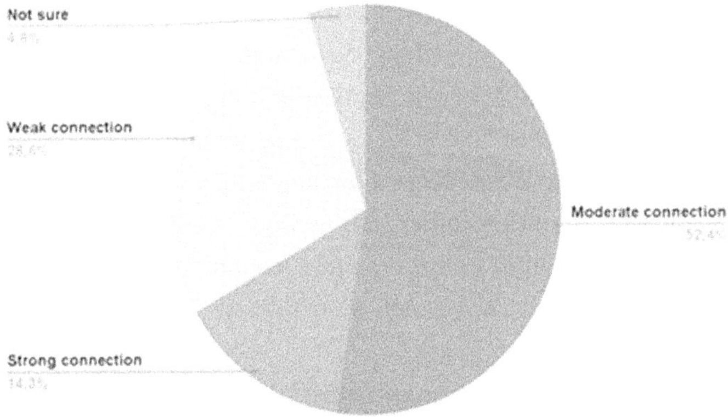

Figure 12.4 Exploring Moroccans' connection to nature. Source: Douae Serraf.

options ranged from "no connection" to "strong connection." The results in Figure 12.4 reveal that the majority of participants, 52.4 percent, reported a moderate connection to the physical environment. A smaller proportion, 28.6 percent, reported a weak connection, while only 14.3 percent felt a strong connection to nature. Some respondents were uncertain and left the question unanswered. Interviewee 25 explained the confusion of the answers and stated: "*How can you be really sure? If it is a matter of whether I go because I am emotionally attached, then no. I rarely go, but I don't mind it.*"

The graph and interviewee's answers showcase the intricacy of determining an individual's connection to the natural world. The interviewee raises a significant point that emotional attachment alone may not be the sole determinant of the strength of someone's connection to nature. Other variables, such as accessibility or personal preferences, may play a role in shaping an individual's bond with the environment. The comment also stresses the subjectivity involved in measuring someone's connection, as what one person perceives as a strong connection may differ from another's viewpoint. In other words, the findings highlight the complexity of measuring the connection between Moroccan individuals and the natural environment.

Moroccans' Emotional Responses to Weather

Participants were then asked a series of questions on their feelings towards different weathers. This question explores the emotions and feelings that people associate with different types of weather. It aims to uncover how people perceive

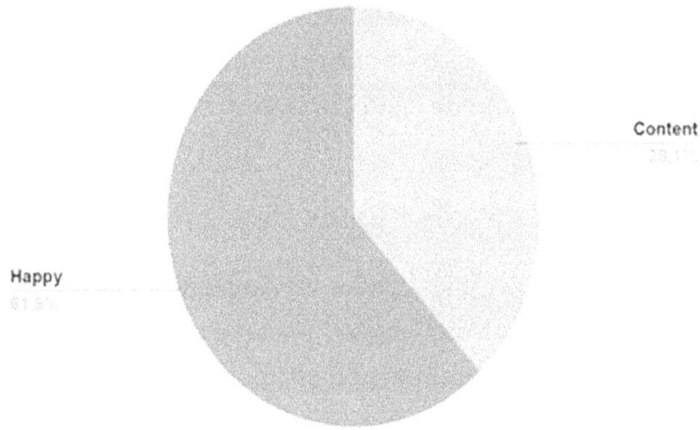

Figure 12.5 Moroccans' feelings toward a Sunny Day. Source: Douae Serraf.

sunny days, rainy days, and days with thunder and lightning. By understanding the emotions that these weather conditions evoke, we can gain insight into the impact that weather has on our daily experiences and moods.

Moroccans' Feelings Towards a Sunny Day

Environmental psychology theories posit that weather significantly influences the physical environment, which subsequently impacts individuals' moods, behaviors, and overall well-being. Studies conducted by Boehm and Lyubomirsky (2009) corroborate that exposure to sunny weather can elicit positive emotions, enhance mental states, and potentially influence behavior. Respondents were presented with five options, along with an additional option to provide a description, ranging from feelings of happiness to feelings of sadness and stress. The responses are summarized in Figure 12.5.

As per the graph, respondents chose two out of the five provided options. A majority of 61.9 percent reported feeling happy during sunny weather, while 38.1 percent felt content. Notably, no participant expressed feelings of sadness, stress, or indifference, thereby validating previous research in environmental psychology. Interviewee 21 described her experiences on sunny days as pleasing and joyful, stating, "*The sun helps me be more outgoing in public; it helps me plan and enjoy my day.*"

Indeed, the findings substantiate that sunny weather can elevate people's spirits, boost their energy levels, and foster feelings of happiness and positivity. This can lead individuals to be more inclined towards participating in outdoor activities, social interactions, and physical exercise. Moreover, sunny weather may motivate individuals to engage in leisure activities such as sports, picnicking, and attending outdoor concerts. Additionally, sunny weather has been observed to positively influence both mental and physical well-being.

Moroccans' Feelings Towards a Rainy Day

Much like a sunny day, individuals exhibit varying attitudes towards rain, which often symbolizes the chill of the weather and the discomfort associated with catching a cold or allergies. Environmental psychology regards rain as an element of the physical environment capable of influencing individuals' behaviors, emotions, and overall well-being. Research in this field indicates that people's perceptions of and responses to rain can be shaped by a multitude of factors, including context, personal differences, and past experiences. Respondents were queried about their feelings during rainfall, and the findings are presented in Figure 12.6.

Results from a survey showed a diverse range of reactions to rainy weather among participants. Nearly 24 percent were annoyed by rain, while an equal proportion felt no particular emotions towards it. 19 percent felt sad during rainy weather, while 4.8 percent reported feeling stressed. On the other hand, 19 percent felt content in rainy weather, and 9.5 percent reported being happy during it. This question provides insight into the diverse range of emotions that people experience in relation to rainy weather. It's interesting to see that nearly half of the participants had mixed feelings towards rain, with 24 percent feeling annoyed and 24 percent feeling indifferent. The remaining participants had

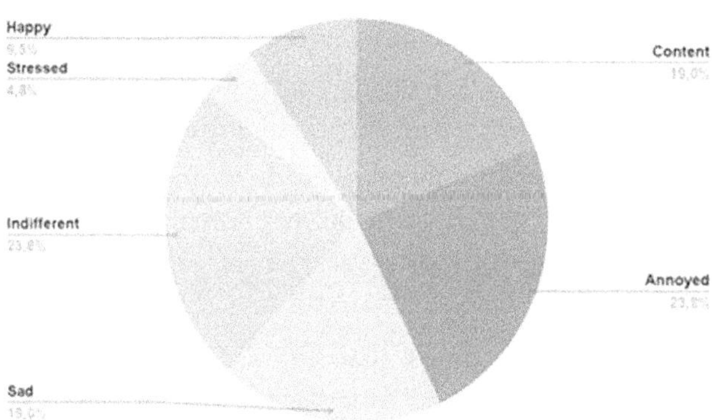

Figure 12.6 Moroccans' feelings toward a rainy day. Source: Douae Serraf.

either positive or negative emotional responses, with 19 percent feeling sad, 4.8 percent feeling stressed, 19 percent feeling content, and 9.5 percent feeling happy. These findings highlight the unique emotional experiences that individuals have with rainy weather, which can be influenced by individual differences, personal experiences, and context.

Interviewee 17 identified rain to be wet. He continues: "*What to like about rain you get wet all the day you get cold; you can't have a good outfit day and worse is when you have to leave your house in the morning.*" On the other hand, interviewee 28 disagrees by saying: "*I am happier with rain that I am with the sun. When I smell the grass and dirt while walking I get a feeling of ecstasy.*"

These differing perspectives highlight the subjective nature of people's experiences with the weather and how individual differences, mindsets, and personality traits can influence their attitudes and emotions towards it.

Respondents' Feelings Towards Thunder

Thunder is considered a part of the physical environment that can affect individuals' mood, behavior, and well-being. Its impact can vary greatly depending on personal experiences and perspectives. While some may see thunder as a frightening and stressful occurrence, with its loud and intense sounds causing feelings of anxiety, others may find it calming and relaxing, providing background noise to help them sleep. Thunderstorms may also bring welcome rain and relief from hot weather, which can positively impact people's emotions. Figure 12.7 presents Moroccans' responses towards thunder. The results indicate the ambiguity of the question. Unlike the sun and rain, people have mixed reactions to thunder. Nearly 48 percent of respondents experience stress, 28.6 percent have negative feelings and are annoyed by thunderstorms, 9.5 percent have positive feelings and are happy, 9.5 percent feel content, and only 4.8 percent are indifferent. Interviewee 10 said: "*The nature of thunder is aggressive. When I hear thunder my heart starts beating fast, and I get scared. I always feel like it would hit me.*" Another interviewee elaborated and said: "*The sound is loud and it doesn't feel. A lot of times my kids get scared when they hear it.*" The data presented highlights the diversity of human emotions and attitudes towards natural phenomena like thunder. A significant number of the respondents hold negative views of thunder, while only a minority have positive or neutral emotions. This emotional variability could be caused by a number of factors, including past experiences, cultural background, or personal

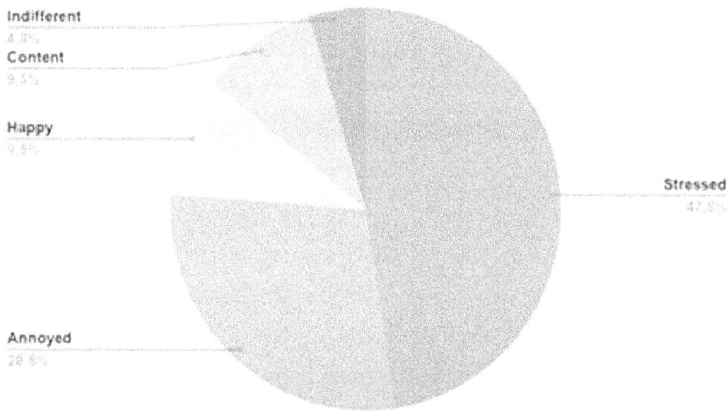

Figure 12.7 Moroccans' feelings toward thunder. Source: Douae Serraf.

disposition. However, these findings emphasize the complexity of this topic and the subjective nature of human emotions and perceptions.

Moroccan Identity in the Physical Environment

This section of the questionnaire aims to investigate the connection between the physical environment in Morocco and the representation of the country's cultural identity. The physical environment encompasses both natural and man-made elements, including landscapes, buildings, and infrastructure. As a reflection of a society's cultural heritage and values, the physical environment can provide valuable insight into a place's cultural identity. This research endeavors to examine the various ways in which the physical environment in Oujda, Casablanca and Tetouan reflects the unique cultural heritage and identity of the country. By exploring the relationship between the physical environment and cultural identity, this study would gain a deeper understanding of the role that the physical environment plays in shaping and preserving cultural heritage. Respondents were asked about the extent to which their Moroccan identity is reflected in the physical environment around them. Figure 12.8 identifies the results.

The graph presents the results of a survey regarding the extent to which participants feel that their Moroccan identity is reflected in the physical environment. 47.6 percent of the participants feel that their identity is somewhat

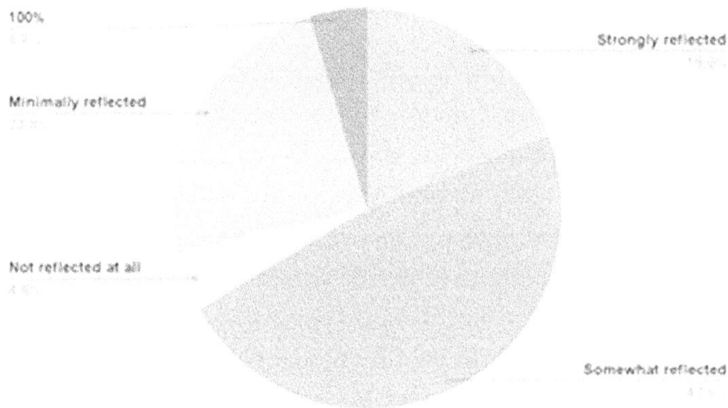

Figure 12.8 Reflection of Moroccan identity in the physical environment. Source: Douae Serraf.

reflected, meaning they believe that there are elements in the environment that capture aspects of the Moroccan identity but not to a significant extent. On the other hand, 28.6 percent disagree and say that the reflection of Moroccan identity in the physical environment is minimal, indicating that they do not see many elements that represent the identity. However, 19 percent of the participants have a more positive view and say that their identity is emphasized in nature, indicating that they see strong representations of the Moroccan identity in the physical environment. Only 4.8 percent of the participants feel that the Moroccan identity is not reflected at all in the environment. In other words, they do not see any elements that capture the essence of their identity. (See Figure 12.8)

Interview 11 discussed the strong reflection by saying: "*Well, it's always a pleasure to look outside and see mountains or trees or Sahara that remind you of your nationality.*" Another interviewee stated: "*As I am from Tetouan, we have the sea and the mountains in our regions which makes it a very pleasing environment. Many countries don't have that. So it is very important for me as a Moroccan.*" These statements show how the physical environment can serve as a symbol of one's identity and bring a sense of belonging and connection to one's culture. Morocco is currently trying to reflect its culture and identity on the built environment and nature to promote tourism and Moroccan heritage. Interviewees had to answer another question regarding the importance of the environment to their sense of identity.

Importance of Physical Environment on Moroccans' Sense of Identity

Interviewee 1 described the importance by saying: "*As a Moroccan, it's a pride to have a green country, it makes our minds greener.*" Another interviewee (25) stated: "*Very important. But the government is not taking it into consideration here near the capital as they only seek money here. The environment is more built for tourists rather than Moroccans themselves and that is horrible for us as a community.*" Although the majority of interviewees agreed with the importance of nature to their sense of identity, others thought it was not so very important to their identities. Interviewee 5 said: "*Well, for the image of my country, maybe but not for my identity.*" These comments from interviewees highlight a range of perspectives on the importance of the physical environment to their sense of identity as Moroccans.

Interviewee 1 values the green environment as a source of pride and views it as a positive influence on their mental well-being. On the other hand, interviewee 25 believes that the government prioritizes tourism over the needs of the local community and this is causing harm to the environment, which is a source of identity for Moroccans. Interviewee 5, on the other hand, believes that while the environment may be important for the image of their country, it is not necessarily important to their individual identity. This highlights that even among the participants who agreed on the importance of the physical environment to their sense of identity; there are varying degrees of significance and opinions on the matter. Another explanation would be some respondents' lack of understanding of the main notions, including identity and the difference between an individual's identity and the one of the country.

Impact of the Physical Environment on Moroccans' Well-being

The physical environment in which a person lives has a significant impact on their overall well-being. This is especially true in Morocco, where the natural landscape, urbanisation, and environmental factors play a crucial role in shaping the daily experiences and perceptions of individuals. In this context, it is essential to understand the relationship between the physical environment and human well-being in Morocco. It is crucial to identify the environmental factors that can contribute to or hinder progress in this area. This includes examining how different aspects of the environment, such as green spaces, air quality, and access to resources, can positively or negatively affect an individual's physical and mental health, happiness, and sense of identity. This section will delve into the impact of the physical environment on human well-being in Morocco and discuss the importance of considering these factors in shaping a sustainable and healthy future for all Moroccans.

The participants were asked about the impact of nature on their moods. Several of them reported having positive experiences in nature. For instance, Interviewee 31 said, "*My mood changes towards the best when I am doing a picnic or when I am round trees with my family.*" Similarly, Interviewee 22 stated, "*Towards the best because when I go to nature I go with my kids and we have a good time.*" However, some interviewees connected nature with calmness and relaxation. Interviewee 8 shared, "*I work in Casablanca, a city almost that never sleeps. Most of the people here don't really care about nature; industrialization is taking place all the time, so my mood is always down. I get really happy to go to cities where the environment is emphasized.* Another interviewee (Interviewee 3) mentioned, "*It changes when it's grey, my mood gets down and I can become sad and I happen to cry from time to time. Winter isn't always fun for me.*" In a similar vein, Interviewee 17 claimed, "*My mood changes with the sea and weather. If there are a lot of movement and grey clouds, I get sad.*"

The participants' responses show a clear correlation between nature and their moods. Those who had positive experiences in nature, such as picnicking or spending time with their families, reported improved moods. On the other hand, those who experienced negative feelings in nature, such as sadness or unhappiness, related it to the sea and weather, specifically grey clouds and a lot of movement. The environment in which they lived also played a role. The statement of interviewee 8 highlights the importance of the physical environment in shaping individuals' well-being and mood.

Impact of Nature on Moroccans' Daily Behaviour

The influence of the natural environment on human behavior is a topic of ongoing discussion and research. This section explores the relationship between individuals and the elements of nature in Morocco. Participants were asked about their daily experiences and how the physical environment, such as weather, nature, and urbanisation, impacts their mood and behavior. This investigation provides insights into the extent to which nature has an impact on the daily lives of individuals and highlights the importance of preserving and promoting a healthy and sustainable environment for all. Interviewees were extremely interactive with this question as it opened the door for more interpretation and description.

Interviewee 23 claimed: "*Yes, when it's gloomy and cold and there is wind, you feel sleepy and want to get home immediately. You wouldn't want to go out or work or talk to people.*" Also, interviewee 6 said: "*Yes, if it rains, I get sad and I don't want to write or work or talk with my family at all. I just lock my room and get under the blanket.*" A third interviewee (15) stated: "*I don't like winter. It gives me allergies and I get depressed and stressed all the time. I don't like to go out most of the time. I become the most unproductive person on earth.*" On the other hand, interviewee 2 shared a different attitude. He said: "*As long as I am breathing well nothing changes my mood or my behaviour during the day.*" The summary of statements shows that not only nature immensely affects humans' behavior, but synchronizes with it.

Discussions

The study found that a significant number of respondents had limited knowledge of the environment, with nearly half confessing to having only a basic understanding. The majority of participants reported a moderate connection to the natural environment, while only a small fraction felt a strong connection to it. The findings suggest the importance of promoting education and awareness about the physical environment and its role in people's lives. Despite the complexity and vastness of nature, having a basic understanding can serve as a foundation for expanding one's knowledge and appreciation of the natural world.

Most respondents and interviewees in the three cities of Morocco found it hard to express themselves about the physical environment. Thinking that nature is something acquired made them fall into the trap of neglect and underestimation. Findings symbolized Moroccans' lack of awareness of nature. Regardless of sex, gender, age, or occupation, respondents tended to choose obvious answers, constructed by the mainstream culture. Whenever they had to think about their connection, impact, and relationship to their identity or behavior, they opted for reductionist descriptions of the physical environment. However, some came to realize that there is an unconscious connection born between them and nature that must be nourished and cared for. Therefore, the synchronisation is undeniable and inevitable. As soon as individuals in Morocco embrace their closeness with the natural environment, the more they will come to appreciate themselves and use that knowledge to accept others and engage with self-esteem.

Conclusion

Environmental psychology in Morocco sheds light on the link between nature and human behavior and its impact on personal identity. Research in this area has demonstrated that exposure to natural environments can bring numerous benefits to individuals, including heightened mood, decreased stress levels, and a stronger sense of self. However, human activities such as urbanisation can also have a detrimental effect on the environment and, subsequently, human well-being. Hence, it is crucial to take into account both the positive and negative consequences of human actions on the environment and work towards preserving and improving the natural environment for the betterment of individuals and communities alike. Therefore, the Moroccan government must adopt more sustainable environmental plans to help citizens engage more in their daily activities and become more productive. Paying attention to the country's image is important, and improving Moroccan tourism is equally essential. However, raising awareness regarding environmental psychology is the most important as humans are a country fuel. In conclusion, as shown in the study, the physical environment, including nature, has a significant impact on Moroccans' sense of identity, well-being, and daily behavior. Many interviewees agreed that the natural environment serves as a source of pride and influences their mental well-being positively. However, there are varying opinions on the degree of importance and the impact of the environment on their identity. The physical environment, specifically the weather and sea, was found to have a direct effect on their moods. The majority of participants reported having positive experiences in nature, while others reported feeling sad or unhappy due to the environment in which they lived. These findings highlight the need for Morocco to prioritize preserving and promoting a healthy and sustainable environment for all its citizens. The physical environment plays a crucial role in shaping the daily experiences and perceptions of individuals, and therefore it is important for Morocco to take steps to ensure that its environment is preserved and promoted for the benefit of its citizens.

References

Aronson, E., Wilson, T. D., Akert, R. M., & Sommers, S. R. (2010). *Social psychology* (8th ed.). Pearson Education.

Altman, I., & Werner, C. M. (1997). *Place attachment.* Plenum Press.

Auclair, L., Alifriqui, M., Simenel, R., & Cheikh, A. (2011). Patrimony for resilience: Evidence from the forest agdal in the Moroccan High Atlas Mountains. *Human Ecology*, 39(4), 405–414. https://www.jstor.org/stable/26268972

Boehm, J. K., & Lyubomirsky, S. (2009). The promise of sustainable happiness. *Social and Personality Psychology Compass*, *3*(1), 1–17.

Brown, G. T. L. (2010). Urbanisation and industrialisation in developing countries: Environmental problems and challenges. *Journal of Environmental Studies*, *40*(2), 77–85.

Canter, D. (1977). The psychological sense of place. *Environment and Behavior*, *9*(2), 171–182.

Fernández-Giménez, M. E., Bathily, A., & Fernández-Giménez, V. (2021). Ilemchane transhumant pastoralists' traditional ecological knowledge and adaptive strategies in Morocco's High Atlas. *Mountain Research and Development*, 41(4), R1–R14. https://doi.org/10.1659/MRD-JOURNAL-D-21-00028.1 bioone.org

Gifford, R. (2002). The role of the physical environment in shaping human behavior, attitudes, and well-being [Electronic version]. *Canadian Psychology*, *43*(2), 123–135.

Gifford, R. (2007). *Environmental psychology: Principles and practice* (3rd ed.). Pearson Education Canada.

Greeno, J. G. (1998). The situativity of knowing, learning, and research. *American Psychologist*, *53*(1), 5.

Jackson, P. (1984). The formation of human identity and the physical environment. In X. Y. (Ed.), *The impact of the physical environment on human identity* (pp. 1–10). Sage Publications.

Kaplan, S. (1983). The role of the physical environment in the hospital. In I. Altman & J. F. Wohlwill (Eds.), *Human behavior and environment: Advances in theory and research* (pp. 187–219). Plenum Press.

Nasar, J. L. (1992). *The evaluative image of the city*. Sage Publications.

Stokols, D. (1996). Translating social ecological theory into design practice: A call for transdisciplinary research. *Environment and Behavior*, *28*(6), 869–897.

Ulrich, R. S. (1984). View through a window may influence recovery from surgery. *Science*, *224*(4647), 420–421.

Werner, C. M. (1997). *Place attachment*. Ashgate Publishing, Ltd.

13

Ecofeminism in Morocco Between Theory and Reality

Hafsa El Bastami

To ensure the proper conduct of research, it is essential to define and explain relevant key terms and theories. Therefore, before introducing the ecofeminist theory, it is crucial to provide a contextual framework on what discrimination against women entails, its link to gender inequality, and the initiation of women's rights culture in Morocco aimed at guiding Moroccan women towards a life free of discrimination. A 2015 report by the National Council for Human Rights discusses four main elements or factors that

> Interact strongly on the particular vulnerability of women and girls to discrimination: poverty, age, disability and social exclusion. Combined, these factors make certain categories of women forgotten from public policies or subordinates. Indeed, during the past decade, the overall improvement in living standards has significantly reduced poverty in Morocco. However, poverty is not only monetary but strongly depends on how laws and institutions function and interact strongly with citizens. (p. 9)

Since its independence in 1956, among the many projects that Morocco has focused on, laying the groundwork for transformations in women's rights has been a priority. Hanafi and Hites (2017) highlight the importance of the work done by Morocco to establish parity between the sexes in constitutional, family, and criminal law. Feminist civil society and non-governmental organizations, composed mainly of educated, urban women, have been at the forefront of organizing and lobbying in favor of these reforms (n.p). ADFM, UAF, AMDF, LDDF, and Joussour, a forum for Moroccan women, were among the first women's organizations in Morocco created in the mid-1980s through the 1990s. These women's organizations were driven by feminist ideology and led the way towards offering a platform for women to speak about their situation, ensuring that women and men are equal before the law, and creating programs and projects dedicated to improving women's economic, social, and political situation.

Gardner (2010) also sheds light on the significance of the projects carried out by the early women's organizations and states that since Morocco gained its independence from France in 1956, there have been many changes to women's rights. Across the nation, women continue to fight for their rights in legal, social, political, and economic contexts. Although work remains, local organizations have made great strides in improving the status of women's rights in Morocco (n.p).

Besides introducing ecofeminism as a theory that examines the links between nature and women and that was coined back in 1974 by French feminist Françoise d'Eaubonne, for the purposes of the research at hand, the beginnings and nuances of the theory of ecofeminism require clarification. The scholarly articles relied on, as indicated in the methods section, are intended to shed light on the development that ecofeminism has experienced. The chapter is divided into three sections: research methodology, results/findings, and discussion, followed by conclusion and recommendations.

Methods

In order to answer the questions meant to compare ecofeminism in theory and ecofeminism in reality, focusing specifically on the case of Morocco, a number of scholarly articles and organization websites' entries have been examined. The purpose was to address the ecofeminist theory and its nuances starting from when it was coined, how it was developed, and whether its theorists faced any hurdles or criticisms along the way. The analysis was done in a form of comparison between what was researched in the literature review, shedding light on few examples of how ecofeminism appears in some countries and how it manifests on Moroccan grounds.

Results

Women and Gender Inequalities

The struggle to eradicate gender inequalities has been ongoing for decades, with a focus on improving women's status at the social, economic, political, and legal levels. Undoubtedly, women in Africa are no exception. They tirelessly contribute to improving their situation. However, several obstacles hinder women's progress and eagerness to reach their full potential. Some of the obstacles that women encounter are described by Marston and Barret (2006) as "lack of access and control over resources. . . . Asymmetrical rights and responsibilities. . . . Women's unique responsibilities and roles provide for less leisure time than men . . . gendered labor markets . . . lack of access to financial markets" (pp. 5–6).

It is important to note that women's lack of access and control over resources is influenced by societal, familial, cultural, legal, and economic aspects embedded in societies. Women's responsibilities and participation in the economic sphere are greatly hindered. Women have many responsibilities that do not allow them to balance between their work within and outside of the household. As emphasized by Marston and Barret (2006): "employment does not absolve women from domestic responsibilities and adds additional demands on their time; thus, when women are employed outside of the household, their overall workload is often increased" (p. 5). Gendered labor markets depicted by sex segregation and wage differences are also a factor. Women find themselves unable to enter the labor market as easily as men; hence, making women seem as objects that are not treated fairly. According to Tackie, Fields, and Siaway (as cited in Boko, Baliamoune-Lutz, and Kimuna, 2005),

> Women in Africa play very important roles in their economies, yet many of them are among the poorest of the population and live on less than $1 a day. Indeed, in many countries, policies to improve economic growth virtually ignore women. (p. 121)

Not only this, but what we can also notice is that women are the category that suffers a great deal from poverty as Kevane (2004) states: "the grinding poverty of African countries falls heavily on the shoulders of women" men and women lead gendered lives because they make different choices, and they make different choices in part because they are presented with different opportunities" (p. 29). This clearly shows the differences that revolve around women and men; differences that always produce positive aspects to men and negative aspects to women. In many places in the world, and not only in Africa, it is highly noticed that patriarchy still plays a crucial role in societies. In Africa for instance as Kevane (2004) states: "African societies are gendered in ways that are detrimental to women's quests for well-being and self-realization" (p. 33).

USAID believes that women are crucial elements in the development process, so the agency is fully committed to offering aid that will help improve the lives of women worldwide as well as working towards achieving women's equality and empowerment. According to USAID: "It is important to engage the untapped energies and abilities of people, especially poor women, if lasting progress is to be made" (USAID website). This clearly means that if women's situation is not improved, then the process of development will not reach its aim and will not be referred to as success. The growth of a nation is aligned with the well-being of women.

Ecofeminism

The articles examined provide a deeper understanding of ecofeminist theory, as well as the nature of the work carried out by organizations in Morocco that aim to improve both women's situation and the environment. According to Cuomo (2002),

> Ecofeminism was originally associated with the view that women and nature are connected in morally significant ways because both are identified with femininity (or traits labelled "feminine"). . . . Ecofeminists therefore took women and nature to be connected because, as providers of life, sustenance, and creativity, they are similarly important and valuable, and their strengths are similarly controlled or violated by men (or "patriarchy"). (p. 1)

Ecofeminist theory draws attention to capitalism and considers it one of the driving factors leading to discrimination against women and the environment. The idea is that patriarchal ideals are so embedded within society that it renders it quite impossible for women and the environment to benefit from it. All that women and the environment experience is the inability to compete with men.

Anjum (2020) also highlights that ecofeminists link women's subordination to the male-dominated culture and add that

> the feminist movement and the environmental movement are seen to work together on the assumption that they both stand for egalitarian, non-hierarchical system . . . in practice, there is enormous evidence historically of women's subordinate position. These inequalities relate particularly to three aspects: 1) the gender division of labour; 2) property rights, especially in land; 3) juridical authority and access to public decision-making forums. (p. 846)

Çelik and Yaşar (2021) argue that ecofeminism positions women and nature at the center of its philosophy, stressing the urgency of a holistic and interconnected worldview. By advancing this orientation and its core principles, ecofeminism aspires to foster critical awareness and a strong commitment to justice that extends across all communities and forms of life.

The common interest shared by theorists of ecofeminism has to do with ensuring that people, regardless of their age, race, gender, and class, live in an environment serving their specific needs while taking into consideration the intertwined powers contributing to it, be it at the political, economic, or ecological levels.

Besides focusing on the connection between women's subordination/nature's destruction and male dominance/patriarchy, ecofeminists want the whole world to understand that women are highly affected by issues attacking the environments since they are less likely to have independence with regards to how resources are managed especially in the cold/hot seasons. Puleo (2017) mentions that: "As several UN world conferences and reports by numerous NGOs point out, women are the first victims of environmental deterioration but also play a key role in the defence of nature" (p. 28). He also adds that: "One of the reasons why ecology has become a feminist issue is that pollution has a particular effect on the health of women and on reproductive health" (p. 31).

Ecofeminist theory has four nuances or dimensions; the first dimension has to do with the theory's link with modern science where the latter is viewed as lacking objectivity and being controlled by men's values. Examples of this are depicted in the marginalization of the contributions of the efforts of the midwives during childbirth while glorifying the more medical aspect of the process.

The second dimension is referred to as vegetarian ecofeminism, which, as suggested by Zein (2017), "puts into practice 'the personal as political', as many ecofeminists believe that 'meat-eating is a form of patriarchal domination . . . that suggests a link between male violence and a meat-based diet'" (p. 3).

The third and fourth dimensions of ecofeminist theory are linked to the materialist and spiritual/cultural views, respectively. The materialist view perceives social hierarchies such as property, power, and labor as the main factors leading to discrimination against women. On the other hand, the spiritual/cultural view, as the name suggests, is about spirituality and the interconnection that people have with the earth/nature.

However, the ecofeminist theory has received a significant amount of criticism, primarily linked to its focus on the idea of dualism. This dualism posits an interconnection between women and nature, and men and culture, without delving deeper into understanding how this particular closeness creates more prejudices between the elements in question. It also limits women and nature to the idea of vulnerability.

The second criticism of ecofeminism is that while mainstream feminism, which is most tightly bound with hegemonic social status, strives to promote equality within the existing social and political structure (such as making it possible for women to occupy positions of power in business, industry, and politics), many ecofeminists oppose active engagement in these areas. These are the very structures that the movement intends to dismantle, as pointed out by Zein (2017).

Ecofeminism and the African Experience

In the African context, Kelleher (2019) posits that Africa is now the "final frontier" for economic models that have already ecologically compromised the rest of the planet. Furthermore, he explains that women and feminist activists are

> already on the front line of the battle for ecological sustainability on the continent. Their [women] everyday struggles, uncompromised commitment, and willingness to envision a radical future in which justice, equity and rights harmonize with environmental sovereignty have the potential to save us all. (n. p)

As this chapter aims to compare ecofeminist theory and its manifestation in reality, it is pertinent to note that Morocco is among the African countries that focus on ensuring gender equality and helping women move out of their vulnerability shell to become more empowered and involved in the projects and opportunities presented to them. Several organizations have been established for this specific purpose.

The Mohammed VI Foundation for the Protection of the Environment is one of the leading institutions that underscore the importance of the role women play in environmental protection. WECF International, a non-profit network, is another example of an institution focused on creating a healthy environment and a gender-bias-free world for all. This network has created training sessions for women, recognizing their contribution to the energy transition and representing opportunities for capacity building and women's economic development. The two major training sessions were held in Tétouan and the Souss-Massa region. There are also other organizations that combine their efforts for environmental good, but do not necessarily place women at the center of their missions and goals.

Discussion

Fighting against women's discrimination has always been one of the major driving factors for women's organizations. Throughout the years, women have experienced different forms of discrimination: on the economic, social, political levels, to name a few. Patriarchal dominance is one of the culprits of such discrimination. Women, however, are not the only ones affected by male dominance; nature is also affected. For that reason, ecofeminists emphasize the fact that women and nature are interconnected, thus creating this idea of dualism. Organizations throughout the world have sought to create opportunities for women that would allow them to not only benefit themselves but the environment as well. Morocco is no exception. In fact, Morocco has always worked diligently to eradicate women's discrimination, and it is showcased in the number of organizations created for this purpose, particularly those that combine it with a focus on improving the environmental situation. Some of these associations are: Association Tahadi pour l'Environnement, Association Akhemmas pour le Développement Rural et l'Environnement, Association Chihab pour le Développement et la Santé et l'Environnement, and Association Ouled Bounaji pour le Développement et la Préservation de l'Environnement (Solovyov, 2021).

However, the question to be asked is whether these organizations are acting from an ecofeminist point of view, with the understanding that women and nature are intertwined, and that when environmental disasters occur, women are on the front line of the most affected entities. As specified by Amili (2022),

> Everyone is affected by climate change in Morocco, but predominantly women and especially women farmers. In rural communities and mountain regions, people live in demanding circumstances and face more profound climate injustice. Women contribute mainly to agricultural production and suffer disproportionately from economic hardships and illiteracy. This makes them more vulnerable to natural disasters and hunger. (n. p)

The African Women Unite Against Destructive Resource Extraction also highlights the fact that women represent the entity that suffers the most from environmental and social devastations: "We are the main victims of oil spills, land grabs, devastated livelihoods, water and air pollution, poor health and high mortality, lack of state services or high-priced commercialised services, and high levels of interpersonal, domestic and sexual violence, often associated with energy resources" (p. 2).

A UN Women article emphasizes that: "Including women and girls in the design and implementation of climate response actions is key to addressing climate adaptation, mitigation and solutions for sustainable development and gender equality" (p. 1). An example highlighting such importance is represented by a group of fisherwomen in Tiguert near Agadir who adopted practices of collecting shellfish that would help support the sustainability of the natural resources for future generations. UN Women's role was focused on supporting the women and putting them

> at the center of any effort to adapt, mitigate and manage risks related to natural disasters while improving their lives and resilience. This includes making a climate-friendly fisheries economy a reality. As part of this work, UN Women has provided fisherwomen with eco-friendly and sanitary equipment for harvesting and for returning juvenile shellfish to their habitats to reduce their loss due to non-selective fishing. In addition, UN Women has also provided the women with neoprene fishing suits with reinforcements in the knees and elbows to protect them from small injuries and help maintain their body temperatures. (p. 2)

The research conducted to see if ecofeminism, the theory, compares to what is actually occurring on Moroccan ground and most specifically through examining the work carried out by the organizations, showcased that there is a lack of literature in this regard; hence, not allowing for a full understanding of how the so-called women's organizations seeking to benefit both women and nature, from their work.

Conclusion

Ecofeminists argue that the interconnection and dualism between women and nature lead to both being discriminated against, albeit in different ways.

Women, lacking independence and control over resources, appear to be the group most affected by the destruction of nature and the environment. However, opportunities have been created to empower them. This empowerment is crucial as it brings women closer to gender equality, a fundamental human right. The conducted research attempted to shed light on the theory of ecofeminism by examining and comparing its aspects with the Moroccan reality. While Morocco appears to be among the leading African countries in creating opportunities and projects to involve women in environmental conservation, the journey is still long and requires more effort. In addition, the scarcity of data addressing women and ecofeminism in Morocco hinders the research process and, in a way, questions the understanding that the women's organizations created for environmental purposes have on the theory of ecofeminism.

Recommendations for future work include conducting fieldwork with institutions and environmental women's organizations to explore the interconnection between their initiatives and the benefits they create for women and the environment. It will also be an opportunity to examine the level of integration that the women have in the decision-making process.

References

Amili, L. (2022). *Women's rights & climate change*. Retrieved from: https://www.equalitynow.org/stories/womens-rights-climate-change-laila-amili-morocco/

Anjum, T. (2020). Ecofeminism: Exploitation of women and nature. *International Journal of English Literature and Social Sciences*, *5*(4), 846–848.

Boko, S. H., Baliamoune-Lutz, M., & Kimuna, S. R. (2005). *Women in African development: The challenge of globalization and liberalization in the 21st century*. Africa World Press, Inc. Eritrea.

Cuomo, C. (2002). On ecofeminist philosophy. *Ethics & the Environment*, *7*(2), 1–11.

Çelik, E., & Yasar, A. (2021). Ecofeminism: Identicalness of women, nature, animals, and food. Retrieved from: https://www.researchgate.net/publication/355031771_ECOFEMINISM_IDENTICALNESS_OF_WOMEN_NATURE_ANIMALS_AND_FOOD

Gardner, S. (2010). The status of women's rights in Morocco. Retrieved from: https://borgenproject.org/womens-rights-in-morocco/.

Hanafi, L., & Hites, D. (2017). *Women's rights in Morocco: Balancing domestic and international law*. Retrieved from: https://www.atlanticcouncil.org/blogs/menasource/women-s-rights-in-morocco-balancing-domestic-and-international-law/

Kelleher, F. (2019). *Why the world needs an African ecofeminist future*. Retreived from https://africanarguments.org/2019/03/why-world-needs-african-ecofeminist-future/

Kevane, M. (2004). *Women and Development in Africa. United States of America.* Lynne Rienner Publishers, Inc.

Marston, A., & Barret, K. N. (2006). *Women in the economy: A review of recent literature.* Retrieved from: http://pdf.usaid.gov/pdf_docs/PNADJ328.pdf

National Council for Human Rights. (2015). *Gender equality and parity in Morocco: Preserving and implementing the aims and objectives of the constitution, executive summary.* National Council for Human Rights, 2015. Retrieved from: https://www.cndh.org.ma/an/node/25012

Puleo, A. H. (2017). *What is ecofeminism?* Retrieved from: https://www.semanticscholar.org/paper/What-is-ecofeminism-Puleo/95f487060f6865039c1c7848187d48a74fe62de9

Solovyov, M. (2021, April 8). *Top 4 Moroccan NGOs making a social impact in women's empowerment.* Retrieved from: https://moroccotravelblog.com/2021/04/08/top-4-moroccan-ngos-making-a-social-impact-in-womens-empowerment/

Zein, L. F. (2017). *General overview of ecofeminism.* Retreived from DOI: 10.31219/osf.oi/fmjqk

14

The Environmental Crisis, Postcolonialism, and Gender Inequality in the Global South

A Postcolonial Ecofeminist Approach

Hajar Taha

One of the most important engines of global economic growth and the eradication of poverty is agriculture. This vital sector accounts for 4 percent of the global gross domestic product (GDP), and in some underdeveloped countries, it can account for more than 25 percent of the GDP (The World Bank, 2022). According to the Food and Agriculture Organization, women in rural areas make up a quarter of the world's population and are subsequently responsible for producing more than a third of the globe's food (2021). In a similar vein, recent studies conducted by the World Food Program suggest that women also comprise up to 40 percent of the world's agricultural workforce and even greater proportions in the least developed countries. In the Global South as a whole, women make up 43 percent of the workforce and make substantial contributions to national food baskets. Women also constitute approximately 66 percent of livestock keepers, and their agricultural activities are substantial for meeting food security. In Sub-Saharan Africa and South Asia, 60 percent of the agricultural workforce is female (2021).

Nevertheless, the sector remains exceedingly underperforming in several countries. This is primarily due to the severe constraints that women, who are usually a salient resource in agricultural production and the rural economy, through their roles as laborers, farmers, and entrepreneurs, regularly face.

As the main custodians of knowledge on crop varieties, natural resource management, and agricultural production, female farmers' activities significantly vary across countries in terms of sustainable resources, land management techniques, and climate resilience. Nonetheless, inadequate access to resources substantially impairs the productivity of at least 500 million female farmers. Gender-specific impediments such as land dispossession, agricultural training, and education put female farmers at a notable disadvantage before they can even sow a seed or plow a field. In Africa, female farmers receive meager compensation for agricultural activities and are sometimes left with no income because of gender inequality, climate change, and their status as post-colonized. In many places across the continent, gender labor division is solidified through deeply ingrained customs and traditions. Fixed gender roles are responsible for restricting women's ventures into new, varied, and often profitable forms of agricultural production. In Morocco, for instance, women were unpaid for their labor before the 1990s. Before that time, women constituted an invisible workforce; thus, "in some regions, such as Haouz, even urban women workers are sought out since women workers are scarce in the countryside due to existing taboos" (Pascon & Ennaji, 1986, p. 58). Landowners are, therefore, keen to hire women because of their cheap labor costs, their presumed obedience, as well as their reputed meticulousness and care in handling fruits and vegetables. These female wage workers are employed without a contract or social security cover and oftentimes suffer from informal working conditions, risking exposure to countless hazards. Female workers in the global supply chain are vital to the global economy. Nonetheless, multinationals compete with each other to reduce production costs to generate a surplus value by lowering labor costs. The result is often jobs that are informal and insecure, involving hazardous workplaces, unpaid overtime, and even forced labor (Bossenbroek, 2021).

There has been little research, however, exploring women farmers' dire situations and challenges in light of their status as post-colonized. A knowledge gap exists concerning the detrimental ramifications of colonial exploitation of female workers in the agricultural sector and the natural environment as well as the effects of women's invisibility on agricultural productivity, efficiency, and sustainability. The aim of this study is, therefore, to address this issue using a postcolonial ecofeminist lens as well as empirical data derived from well-documented studies. In this respect, the overall structure of the study takes the form of six chapters, including this introductory chapter. Chapter one begins by laying out the theoretical dimensions of the research and looks at how postcolonial ecofeminism supports small-scale female farmers and promotes sustainable agroecology. I then investigate how women's momentous contribution to agricultural productivity is systematically overlooked and explore gender and climate bias. This knowledge is critical because, in 2015, over

98 percent of the world's 795 million hungry people lived in the Global South (McGuire, 2015). The aim of this study is also to demonstrate that women have been subjected to colonial exploitation of assets and labor, continue to endure post-colonial damages, and are disproportionately more susceptible to climate vulnerability. Finally, I conclude that dismantling gender-specific and climate-bias obstacles in agriculture would not only assist women to achieve their highest economic potential but will also significantly contribute to the achievement of global food security.

Postcolonial ecofeminism was chosen as a background theory because it challenges the predominant consideration of anthropocentrism and patriarchy as the archetypal loci of anti-ecological thinking. Postcolonial ecofeminists believe that colonialism is more responsible for the destruction of women's productivity, both by removing the land, water, and forests from their management and control, as well as through the ecological destruction of these resources (Hay, 2002). Women in rural areas frequently face unequal access to resources and limited mobility, compromising their capacity to provide for themselves and their families. Women and children are particularly vulnerable to natural disasters such as climate change, extreme droughts, and floods because they possess fewer rights and economic resources. Postcolonial ecofeminism also provides a broader framework for investigating female farmers' interdependence with the natural environment. Examining third-world women's long history of colonial oppression offers particular insights to postcolonial ecofeminists, who examine the special relationship between women's lived experiences and nature to build a social and political movement. Ecofeminism is therefore less adequate to interpret Third World women's relationship with their natural environments. That is, other than the works of Shiva (1988), Merchant (1989), and Agarwal (1991), seldom has ecofeminist study explored the relationships between rural women's experiences and the natural world. Mostly, states Carolyn E. Sachs, "women's relationship with the natural world is described in metaphors associating women with nature or in the search for past matrilineal cultures that both granted relatively high status to women and had harmonious relations with the natural world" (Sachs, 2019).

Concerning terminology, the Global South refers to the countries that are mostly located in the southern hemisphere. These countries are characterized by an alarming poverty rate and are subsequently more economically dependent on exporting primary products. It should also be noted that many of the countries in the Global South were former colonies of countries in the global North.

Methods

In this study, I use a postcolonial ecofeminist lens to dissect the multifaceted oppression experienced by female farmers in the Global South. Theoretical concepts like the "feminization of nature," "female/male" dichotomies, and "invisibility of women's labor" are thus used throughout the work. I also base my conclusions on findings from well-documented empirical studies. Furthermore, in order to avoid generalization and strike a balance between on-the-ground women's experiences and broader studies of global and regional trends, this study employs a level-based analytic structure. As a result, the research is carried out at three levels:

- At the global level, incorporating female farmers from the Global South
- At the regional level, investigating female agriculture in Africa
- At the local level, as illustrated by Morocco

Africa was chosen as the Global South's representative because women play a significant and critical role in the global agricultural labor force. Women do the majority of the work in food processing, production, and marketing. They are at the cutting edge of agricultural production. Nonetheless, women are significantly underrepresented in agricultural research when it comes to establishing priorities, shaping research agendas, decision-making, and governance. On a local level, Morocco was chosen because agriculture is one of the country's main economic drivers. The sector accounts for 15 percent of the country's GDP on average and employs roughly half of the labor force. Agriculture accounts for 23 percent of the country's exports (Morocco - Agricultural Sector, n.d.). Morocco and other African countries share some historical and cultural contexts regarding the gender division of labor. Moroccan female farmers, like those in many other African countries, continue to have disproportionately low levels of land ownership, operational holdings, and decent working conditions. They are also more at risk from climate change and other environmental issues.

The approach employed in this study includes collecting and contextualizing research to derive conclusions and analyze issues from the survey itself, as well as providing a solid foundation for identifying the multifaceted oppression that women in the Global South are particularly vulnerable to. The research is presented in sections that begin with a discussion of postcolonial ecofeminism's pivotal role in championing the cause of 'postcolonial' female farmers of the Global South. Additionally, before discussing climatic influences on productivity and food security, as well as biases and obstacles to women's farming, land retention, and acquisition, there is a section that details women's agricultural production and precarious conditions. Thereafter, broader conclusions and results are generated and addressed across various topics, preceded by immediate remarks on gender problems, justice, and efficiency effects, as well as suggestions for eliminating the stereotypes that inhibit women's agricultural output.

Ecofeminism, Postcolonialism, and Female Farmers

In recent years, many academic canons have undergone critical scrutiny. Environmental criticism, deep ecology, and feminism have all called for a reassessment of the Western Cartesian mentality (Moyer, 2001). Ecofeminism, which is based on the premise that the domination of women and nature are conceptually interlinked, has also solicited a re-examination of the Western philosophical heritage. In this regard, men's neglect and oppression of women and the environment are explained, according to ecofeminism, by the male understanding that characterizes the worth of women and nature in relation to itself. As a reaction to this androcentric patriarchal perspective, ecofeminists prefer to create a harmonious symbiosis with the nonhuman environment and respect it for its inherent worth rather than its usefulness for humans.

Ecofeminism was coined by the French feminist Francoise d'Eaubonne, who called upon women to lead an ecological revolution to save the planet" (Merchant, 1995 p. 184). D'Eaubonne sought to describe the unspeakable violence inflicted upon women and nature as a result of male domination. Therefore, ecofeminism is the social movement that considers the oppression of women and the environment as interrelated. Consequently, the term implies a link between the home and the earth. Women form this relationship in order to maintain their homes and the Earth (Sandilands, 1997). Ecofeminism is more than a term; it is a framework for combining, re-examining, and enhancing the environmental and feminist movements.

It is important to note that ecofeminism is an umbrella term that encompasses a plurality of ideas and strongly denounces the system of capitalist patriarchy. That is why it cannot be restricted to a predetermined border or explained with a single definition. Ecofeminists describe ecofeminism as a natural ideology that relates women to nature and highlights the function of every living creature in the biosphere. They condemn institutionalized social practices which justify the domination of males over women and promote placid harmony and ecological awareness among all living organisms on the planet. Furthermore, ecofeminism challenges the logic of dominance based on oppositional value dualisms such as men and women, nature and culture attributing more value to men and culture and considering women, animals and plants as means to a profitable end. In this regard, the female-centered ecophilosophy equates the subordination of nature to the persecution of women and criticizes the dominant androcentric attitude. Similarly, ecofeminism criticizes the dominant conceptions that associate women with nature owing to their common "feminine" characteristics and males with culture due to their mutual 'masculine' characteristics (Warren, 1990).

Associated fundamentally with ecofeminism and postcolonialism, postcolonial ecofeminism's basic assertion is that the concomitant deliverance of nature is conceptually interlinked with the liberation of women of color, working-class women, and Third World women. Challenging the predominant consideration of androcentrism and patriarchy as the archetypal locus of anti-ecological thinking, postcolonial ecofeminists believe that colonialism is responsible for the degradation of 'women's production by taking the land, water, and forests from their administration and control and by destroying these resources ecologically (Hay, 2002). Women in rural regions frequently face uneven access to resources and limited mobility, limiting their capacity to sustain themselves and their children.

Postcolonial ecofeminism is a theoretical framework that asserts that Western feminism is insufficient to advocate for all women worldwide. It combines the interests of postcolonial ecocriticism and postcolonial feminism to argue that the material conditions of women in the Global South are the primary determinants of their relationship with their environment. Mukherjee's (2010) definition of postcolonialism as "intensified and sustained exploitation of the human and the non-human" emphasizes the assumption that the material conditions of (post) colonized individuals, men and women, are immediate consequences of their colonial position.

Proponents of postcolonial ecofeminists argue that the material conditions of women in the Global South are the primary determining factors of their relationship with their environment. They contend that women are the primary victims of environmental degradation since they are more dependent than men on agriculture and forest products. Furthermore, they challenge the logic of domination arising from oppositional value dualisms such as men and women, nature and culture, which attribute more value to men and culture and consider women, animals, and plants as means to a profitable end.

Equally important, postcolonial ecofeminism encompasses a wide array of perspectives, such as toxic waste, deforestation, reproductive rights, and local and international agricultural development. Ecofeminists argue that a fair and realistic depiction of women's significant involvement in agricultural production should start with some revisionist history. For instance, Autumn Stanley (1981) states in a survey of anthropological and mythical sources that "from the earliest ages until at least the horticultural period, women were principally responsible for the gathering, processing, storage, and finally for the production of plant sustenance." Postcolonial ecofeminists also believe that women are the primary sufferers of environmental degradation since they are more dependent than men on agriculture and forest products. As a result, they are more susceptible to environmental crises such as drought, climate change, and forest resource depletion as their source of income usually comes from the products that they make from farming.

In this regard, postcolonial ecofeminists argue that despite women's significant contribution to agricultural production in third-world countries (women constitute between one-third and one-half of the agricultural laborers in the Third World), the gender division of labor typically allocates tasks such as cash crops to men, while the labor-intensive chores of food crops are usually relegated to women (Merchant, 1995). For instance, women in Africa are known to produce more than 70 percent of Africa's food, often without the aid of tractors, oxen, or even plows (Warren & Perlez, 2010). Furthermore, women comprise 43 percent of the world's agricultural labor force, a figure that rises to 70 percent in some countries. In this regard, Buvinić and Yudelman (1989) say that

> As a rule, women farmers who work longer hours have fewer assets and lower incomes than men farmers do and have almost as many dependents to support. The disparity is not due to a lack of education or competence; women farmers are poorer because their access to credit is limited. Without credit, they cannot acquire productive assets, such as cattle, fertilizer, or improved seeds, to improve the productivity of their labor. (p. 30)

As stated above, this paper argues that Western ecofeminism falls short in interpreting the relationship between Third World women and their natural environments. This is partly due to its neglect of the historical and material conditions of Third World women and the tendency to view "Third World woman" as a homogeneous category. This generalization fosters a discourse of "otherness," where "Third World women" are compared against women of the First World. Therefore, it is crucial to recognize the position of Third World women as (post)colonized to analyze their complex and ambiguous attitudes toward their natural environments.

Invisibility of a Female Agricultural Workforce

Women constitute 40 percent of the agricultural labor force globally and are critical to the nutritional well-being of their families and communities. However, many female farmers in the Global South are still looked down upon as simple workers with little to no decision-making authority, rather than as autonomous farmers seeking equal opportunities to those granted to men. This means that the gender agricultural productivity gap is still alarmingly persistent and greatly hampers women's productivity. Without land ownership, women are left with nothing to leverage as collateral for formal finance and are often precluded from government programs to support farmers. Ultimately, female farmers' limited access to documented land rights increases their 'invisibility' and limits their capacity.

In Africa, a multi-country study has reported that the gender gap in productivity ranged from 13 percent in Uganda to 25 percent in Malawi, even though women had access to land, partly because women lack secure land tenure and have limited access to education, formal training, and markets than men (Velde et al., 2020).

Although female farmers' income is critical to the survival of their families, their invaluable contribution continues to be invisible and stigmatized in society. Women are exceedingly involved in domestic activities, including cooking, cleaning, and raising children, which remain economically invisible. The pre-established gender roles that consider men as the sole providers for their families and women as housewives remain pervasive in rural areas. This image is further promoted by most employers, who often hire female farmers without employment contracts. Accordingly, colonial discourses advocating the "housewifization" of women and female domestication are at the heart of reconstructing patriarchal relations in rural areas. Domestic discourses concerning rural families not only reinforce patriarchal relations but also continue to legitimize the subordination of female farmers through romanticized narratives that deliberately undermine rural women's contributions (Sachs, 2019). Ultimately, this refusal to recognize women as the main earners of livelihood systematically sanctions the employers' stigmatization of female workers. Women are in most cases given low wages, often 20–40 percent less than their male counterparts. Employers also justify this form of segregation in terms of the specific activities that women undertake, which supposedly need "less physical strength" (Bossenbroek, 2021).

Another point worth considering is that the complex and ambivalent relationship of women in the Global South with the land is fundamentally affected by their status as (post)colonized. Studies of female farmers' work during the colonial period provide compelling evidence that they lost control over their land as well as their economic autonomy with the arrival of cash crops and women's institutionalized exclusion from the global marketplace. Not to mention, men and international commerce were able to make substantial profits because they relied to some extent on women's unremunerated labor (Sheldon, 2013). According to Carolyn E. Sachs, colonial agents and officials used arguments astonishingly similar, in southern and eastern Africa, to those used to persuade Native Americans to make radical social reforms during the 1900s as they endeavored to push men into farming. As a direct outcome of these colonial policies, colonial and postcolonial governments in eastern and southern Africa, in a similar fashion to South Asia and North America, succeeded in convincing Africans to consolidate land in men's names, with the objective of enhancing cash crop production. Ultimately, the end of the colonial era in Latin America, Asia, and Africa marked the initiation of multiple types of land reform, including the privatization of common lands, land resettlement schemes, and the establishment of communal farms, which were initiated to suppress peasant resistance movements. Nevertheless, land reforms that were developed to stifle rebellions ended up hampering women's access to land.

In Morocco, for instance, the French government's colonial agenda was to transform Morocco into its "granary" while developing local farmers' production. This equation came with many deleterious ramifications both on the land and on women's situation. On the product distribution side, French suppliers exerted a monopoly on access to global markets, preventing local suppliers, particularly women, from capturing part of the trade rent. On the land distribution side, the collective land ownership act issued by the French authorities forbade private appropriation, while giving the government the discretionary power to allocate vacant plots to settlers. This, in turn, set the stage for the expropriation of large plots against farm holders' consent and sometimes with the use of force. Eventually, the adoption of land extensive technics, soil exhaustion, and labor drain by colonial settlement not only resulted in a loss in local productivity but also imposed land dispossession on local female farmers as well as resource deprivation (Salem, 2021). Increasing privatization later forced women to either lose access to land or to land rights because individual ownership of land evidently went to men. The real tragedy is that colonial policies denied women their basic rights to land ownership and laid the foundation for the contemporary forms of women's subordination (Whitehead & Tsikata, 2003).

Aside from the formal and informal barriers to accessing and owning land that women face in the Global South, women's traditional knowledge of agriculture is also criminally disregarded. Traditional knowledge in agriculture refers to the accumulative body of practices, customs, and beliefs that play a vital role in the maintenance of land, water, and agricultural resources. It is also accounted for in the decision-making process of various stages of production, distribution, and consumption of almost every agricultural resource (Berkes et al., 2007). When it comes to women's traditional expertise particularly, women are highly considered as the gatekeepers of conventional natural resource management knowledge. Rural women are the custodians of conventional natural resource management knowledge. Their dependence on natural resources to provide daily subsistence for their families is the reason why they possess strong traditional ecological knowledge and a keen interest in environmental protection and management. A recent study conducted by the African Biodiversity Network, the African Women's Development and Fund, and the Gaia Foundation found that small-scale farmers, mainly women, are responsible for the production of 80 percent of the food in Africa on only 14.7 percent of the agricultural land. These farmers also control 80 percent of the seeds produced and exchanged on small farms. The study offers insight into the intricacy of women's agricultural expertise and their inherited awareness of the nutritional and cultural demands of the community which are at the core of food sovereignty (Courtine, 2018).

Additionally, women nurture a peculiar symbiosis with the environment and tend to be more concerned about the availability and access to natural resources. In this regard, renowned ecofeminist Howard (2003) argues that

> Women predominate in plant biodiversity management in their roles as housewives, plant gatherers, home gardeners, herbalists, seed custodians, and informal plant breeders. But because most plant use, management, and

> conservation occur within the domestic realm, and because the principal values of plant genetic resources are localized and non-monetary, they are largely invisible to outsiders and are easily undervalued. (p. 3)

Similarly, postcolonial ecofeminist Shiva (2016) draws a clear distinction between men's and women's agriculture and agricultural knowledge. Conventionally, she notes, "the backyard of each rural home was a nursery, and each peasant woman thc silviculturist. The invisible, decentered agroforestry model was significant because the humblest of species and the smallest of people could participate in it" (p. 79). She additionally argues that this knowledge was replaced, in masculinist development projects, by the "reductionist mind" of the experts who willingly overlooked the multiple uses of traditional planting and its use. Along the same line of arguments, Deane Curtin asserts that the most powerful tool of invisibility and marginalization in the case of women's agricultural knowledge is the "culture/nature" and the "male/female" dichotomies. According to him, while the perception of these dichotomies varies dramatically within different cultures, a common ground between all cultures resides in their ongoing dependence on the masculinist model of development. The masculinist standpoint in these dualisms inherently assigns a higher value to the "modern," rational scientific, high-tech, and global over the traditional, low-tech, and small scale. Men are construed by the same dualisms as autonomous, rational, and independent, whereas women are considered dependent or interdependent. Subsequently, men's agriculture is construed as a theoretical practice or an applied science while women's expert knowledge of seeds, soil, and climate is ridiculed as anecdotal and dismissed as "old wives' tales" (Warren et al., 1997).

It is important to note that postcolonial ecofeminists consider female agriculture an ecofeminist issue. Agriculture relies on the biodiversity of an ecosystem that can supply fuel, fodder, medicine, and food. Caring labor is, as far as postcolonial ecofeminists are concerned, highly indispensable for the holding together of the family as well as the environment. Furthermore, ecofeminists believe that small farmers and biodiversity constitute the foundation of food security. They also call for the preservation of their indigenous knowledge of the sustainable ecological community (Warren et al., 1997). In Morocco, for instance, women's traditional knowledge is included in the creation of natural products obtained from natural resources. The government is very supportive of and accords an increased significance to policies that support the establishment of income-generating enterprises through cooperatives. The current trend of traditional knowledge promotion in women's cooperatives is believed to foster social and financial autonomy and remedy the existing gender gap in agricultural production, especially in the context of the Green Morocco Plan. The Green Morocco Plan (2008–20) was officially designed to address the country's most poignant issues of climate change, natural resource conservation, poverty alleviation, economic growth, and food security. Additionally, it attempts to improve the living conditions of women in rural areas by addressing the challenges to women's employment and social integration. Nevertheless, women are still subjected to the same facets of oppression that continue to hinder their social, economic, and financial evolution (Montanari& Bergh, 2019).

Climate and Gender Bias Impacts on Female Farmers' Productivity and Food Security in the Global South

Melting glaciers, changing weather patterns, intensifying climate disasters, droughts, and floods are exceedingly devastating crops, disrupting rural communities, and damaging livelihoods dependent largely on the agricultural sector. Agriculture is very sensitive to weather and climate because it depends heavily on water, land, and other natural resources that climate affects (Walsh, 2020). Disruptions in weather patterns can negatively affect food availability and quality and reduce access to food. Agricultural workers also face various climate-related risks. In the Global South, women play a crucial role in the agricultural sector and are subsequently affected by climate change and other poignant environmental issues due to the same barriers that reduce their productivity. Gender bias is one of the direct factors in women's suffering from the adverse effects of climate change as economic, social, and political barriers further exacerbate women's susceptibility to the detrimental impacts of climate change. Subsequently, a gender-conscious approach to climate change resilience in the agricultural sector is increasingly pivotal to addressing the high levels of climate vulnerability that women are exceptionally subject to.

Globally, in a violation of gender and distributive justice, women risk hunger 10 percent higher than men. Women in the Global South are particularly susceptible to hunger despite their agricultural work. Africa, for instance, is poised, according to the latest report of the Intergovernmental Panel on Climate Change (IPCC) to be among the most vulnerable regions to climate variability and unpredictable weather patterns. According to the same report, climate vulnerability in Africa has witnessed an alarming increase over the past 50 to 100 years, with detrimental effects on food security, livelihoods, and the health of people. Furthermore, climate change is expected to increase water scarcity, lessen crop yields, and aggravate biodiversity loss, thus posing a severe risk to the continent (IPCC, 2022). The African continent has also been experiencing multiple humanitarian crises in hunger. This situation is further aggravated by multiple developmental challenges such as complex governance, endemic poverty, climate injustice, and institutional dimensions. Correspondingly, climate change has a major impact on agricultural productivity.

In Morocco, the detrimental impacts of the climate crisis are increasingly alarming. Morocco has been identified, by the 4th Assessment Report of the United Nations Intergovernmental Panel on Climate Change, as a very vulnerable country. Various regions across the country are exceedingly affected by an increase in average temperatures, drought, heat waves, extreme rainfalls, and sea level rise. Along with the changing weather patterns, the mountain snow cover is expected to significantly decrease. Annual rainfall is also dropping at an alarming rate and affecting large parts of the country. To make matters worse, projections indicate a drop in average precipitations of 10 to 20 percent across the country, with a 30 percent decrease for the Saharan region by 2100 (3rd NC Morocco, 2016). This setting already has significant impacts on agricultural production, water resources, and the whole vegetation of the country. As a rule, rural women are expected to be more exposed to poverty and unemployment. This population is largely responsible for agricultural labor, cultivation, and maintenance of the various orchards. Rural women also walk long distances to draw and transport water home. Furthermore, they live fundamentally on the exploitation of natural resources (animal husbandry, biomass energy) and are totally dependent on climate change (Climate Expert: Morocco, n.d.) Ultimately, these women's source of income is significantly dwindling as they remain powerless in the face of extreme weather patterns as well as the waves of migration of their children who leave their villages towards other regions in search of a better life.

Colonialism has also played its part in exacerbating the climate crisis in the Global South. According to postdoctoral scholar Hadeel Assali, "Western colonial legacies operate within a paradigm that assumes they can extract its natural resources as much as they want, and the Earth will regenerate itself" (Varanasi, 2022). This claim was supported by the Intergovernmental Panel on Climate Change (IPCC), which considered colonialism a major driver of the climate crisis for the first time since its inception. Current development challenges causing climate vulnerability are exceedingly impacted by historical and ongoing patterns of political injustice such as colonialism, especially for many Indigenous peoples and local communities.

A recent study published in Nature suggested that research investigating climate change impact in the Global south is less likely to be conducted. It also highlights the fact that these seventy-eight low-income countries are facing the worst ramifications of climate variability. Additionally, these countries are also the least responsible for greenhouse gas emissions; Africa in its totality accounts for the lowest share of climate-warming emissions at 3.8 percent. Conversely, major industrial countries such as the US and European Union are responsible for 19 percent and 13 percent of global emissions (Callaghan et al., 2021).

Ongoing colonial conservation efforts also affected indigenous communities, which ended up losing access to their lands as well as their primary source of income from agricultural production. Today, unjust climate policies continue to exclude indigenous people, especially women, despite their disproportionate susceptibility to extreme weather events.

In fact, acknowledging female farmers' status as 'post-colonized' is the first step towards improving their situations. In the long run, providing female farmers with various opportunities to contribute to climate-smart agricultural programs and including them in the various steps of decision-making concerning agricultural activities has been proven to significantly improve yields and climate resilience. As a matter of fact, facilitating rural women's access to productive resources and inputs is expected to increase their productivity by twenty to 30 percent. This, in turn, is expected to yield various fruitful results such as an increase in agricultural outputs in developing countries by 2.5 to 4 percent (UN Women, 2020). In this regard, the IPCC (2014) makes the following recommendations:

- Adaptations in technology, such as stress-tolerant crop types and irrigation;
- Alternative sources of income and diverse occupations;
- Access services for small-case farmers to bank accounts and financing in order to get access to production materials and agricultural supplies;
- Government support for the development of early-warning systems for weather and other environmental risks. Government support for the development of early-warning systems for weather and other environmental risks; and
- Adjustments in agroforestry and sustainable agriculture in terms of agronomy. (p. 2)

Finally, women have been universally acclaimed as the backbone of agricultural production, the primary guardians of agricultural knowledge, and the main custodians of food security. Subsequently, they should be granted access to productive resources and inputs that could increase their productivity and agricultural outputs in developing countries and reduce the number of hungry people in the world by 12 to 17 percent, or 100 to 150 million (UN Women, 2020).

Findings and Discussion

The results of this study indicate in section 3 that postcolonialism is indeed more suitable as a background theory to investigate the lived experience of female farmers in the Global South. Postcolonial ecofeminism adopts a gender-specific perspective to transcend whitewashing and consider how the experiences of gender-variant individuals and third-world women face structural injustices owing to the existence of such oppressive frameworks as patriarchy, sexism, and the long-term consequences of colonialism. The agricultural sector in postcolonial countries is a particularly rich context to reveal the interconnectedness of ecological deterioration and highlight the oppression based on gender. The findings also confirm postcolonial ecofeminism's basic premise, which suggests that the ideas originating in the Global North cannot simply be used to interpret the experiences of the rest of the world. Instead, it attempts to incorporate into the global ecofeminist sphere the perspectives and experiences of indigenous communities and women from across the Global South.

Section 4 provides evidence that the commercialization and capitalization of agriculture is a direct outcome of colonialism. These practices not only contribute to women's declining control of and access to land but also have detrimental effects on the environment. Women were forced, under colonial rule, to grow export commodities that yielded meager financial gains. Subsequently, they were coerced into neglecting their own food crops and faced major issues of food shortages. The introduction of cash crops by colonial agents also drastically changed women's economic situations and their control over land. Patriarchal organization of rural households and colonial ideologies deeply ingrained in African countries continue to erect major barriers to women's access to land. Prevented from owning land, female farmers are left with no source of collateral, which means that they are denied loans to buy more land and adequate equipment to keep pace with the latest technological advances in agriculture. In most areas, women continue their work growing food and working on the family farms while men earn wages and benefit from financial support.

Section 4 also provides evidence that women's traditional knowledge transfer is a key step toward product development and is key to the production process itself. It also shows how women's productive work remains unrecognized and continues to be devalued. Third-world women are oppressed by both international and national injustices and by social institutions that grant husbands, fathers, and brothers absolute dominion over women. The unjust distribution of power, resources, and responsibilities is sanctioned by the larger system of capitalist patriarchy, legitimized by ancient traditions, and ingrained into women's own attitudes. Subsequently, it is essential to understand and then modify the societal conventions that constrain women's prospects. By adopting a gender transformational strategy, we may permanently close the gap in access to resources and services between men and women.

Section 5 suggests that rural women play a pivotal role in agricultural development in many developing countries. The hardships faced by female farmers in the Global South heighten their susceptibility to climatic and meteorological circumstances, diminish their resilience, and hamper their capability. The section also provides evidence that colonialism is a historic and ongoing driver of the climate crisis. The devastating impacts of climate change on agriculture are also key factors in the deterioration of women's economic situation. Women suffer from the most precarious of conditions and work under perilous situations in the agricultural sector. They also constitute the majority of the world's poor. Their adaptability to shifting agricultural opportunities and constraints in a changing climate is thus highly restricted. That is why the empowerment of female agricultural workers should be regarded as an essential step toward poverty reduction, economic advancement, and sustainable climate adaptation strategies.

Conclusion

In a time of unprecedented human footprints disturbing the earth's ecology across the globe, postcolonial ecofeminism calls for the adoption of a feminist perspective in a green-centered society. It highlights the urgent need for a collaborative, egalitarian community whereby human beings live in symbiosis with other constituents of the biosphere. Postcolonial ecofeminist analysis both unearths the harmonious interconnectedness between women and nature and addresses the parallels between their mutual subordination. Subsequently, female agricultural production is one of its main themes. Women farmers play a major role in small-scale agriculture, which is more effective at reducing hunger and poverty than any other sector. They are also responsible for the production of a significant amount of food globally. Nevertheless, few measures have been taken to guarantee their access to resources, improve their living conditions, and value their contribution to their community's resilience to climate change. Women are dramatically underrepresented within policy and decision-making related to agriculture, and their agricultural work still lacks formal recognition.

Female agriculture workers are also the first victims of environmental calamities because they constitute the majority of the poor, and their livelihoods largely depend on natural resources. They also face significant social, political, and economic obstacles that limit their coping capacities and resilience. In other words, female farmers are the first to suffer when environmental catastrophes occur and the last to gain when economic profit takes place.

Women in the Global South have also suffered because of the long periods of colonialism that their countries witnessed. Land tenure policies were at the heart of women's modern-day disposition. Women are less autonomous because the introduction of cash crops in the colonial period resulted in their steadfast exclusion from the global marketplace. However, all is not lost if we work to mobilize our efforts to introduce significant changes in agribusiness. Female farmers should, therefore, be granted access to modern, labor-saving technologies. Their invaluable contributions should be recognized internationally, and specific legal regimes should be established to ensure they can benefit from their own knowledge and prevent its illegal use by others. The integration of women's agricultural knowledge into local education systems would provide much space for knowledge-transfer opportunities. Such an increase in both participation and recognition can significantly contribute to widespread social and political changes in the Global South and enhance sustainable development, food security, and biodiversity. Professor Ruth Oniang'o relatedly notes, "with the right support, women can rid Africa of hunger, malnutrition, and poverty" (World Bank Group, 2018).

The following conclusions may, therefore, be derived from the current research and need additional examination. First, female farmers in the Global South face a variety of gender prejudices that significantly reduce their efficiency.

Second, one of the immediate consequences of colonialism is the subjection of women. Third, the traditional knowledge of female farmers is more likely eco-sustainable than modern practices in industrial farming. Finally, women's food security and well-being in the Global South are in a humanitarian crisis, which has been aggravated by the climate disaster. Each of these factors needs more examination to enhance the lives of female farmers, their communities, and future generations. Future research on the present issue is thus strongly suggested.

References

Agarwal, B. (1991). *Engendering the environment debate: Lessons from the Indian subcontinent* (CASID Distinguished Speaker Series, Discussion Paper No. 8). Center for Advanced Study of International Development, Michigan State University.

Berkes, F., Berkes, M. K., & Fast, H. (2007). Collaborative integrated management in Canada's North: The role of local and traditional knowledge and community-based monitoring. *Coastal Management, 35*(1), 143–162.

Bossenbroek, L. (2021, October 22). *Beyond Moroccans agricultural boom: The invisibility and precariousness of a female agricultural workforce*. Arab Reform Initiative. Retrieved from: https://www.arab-reform.net/publication/beyond-moroccos-agricultural-boom-the-invisibility-and-precariousness-of-a-female-agricultural-workforce/

Buvinić, M., & Yudelman, S. W. (1989). *Women, poverty, and progress in the third world.* Foreign Policy Association.

Callaghan, M., Schleussner, C. F., Nath, S., Lejeune, Q., Knutson, T. R., Reichstein, M., Hansen, G., Theokritoff, E., Andrijevic, M., Brecha, R. J., Hegarty, M., Jones, C., Lee, K., Lucas, A., van Maanen, N., Menke, I., Pfleiderer, P., Yesil, B., & Minx, J. C. (2021, October 11). Machine-learning-based evidence and attribution mapping of 100,000 climate impact studies. *Nature Climate Change, 11*(11), 966–972.

Climate Expert: Morocco. (n.d.). Retrieved from: https://www.climate-expert.org/en/home/business-adaptation/morocco/

Courtine, E. (2018, October 9). *Celebrating African rural women: Custodians of seed, food & traditional knowledge for climate change resilience*. ICCA Consortium. Retrieved from: https://www.iccaconsortium.org/index.php/2016/02/29/celebrating-african-rural-women-custodians-of-seed-food-traditional-knowledge-for-climate-change-resilience/

Food and Agriculture Organization of the United Nations. (2021, April 23). *Small family farmers produce a third of the world's food*. FAO Newsroom. https://

www.fao.org/newsroom/detail/Small-family-farmers-produce-a-third-of-the-world-s-food/en

Hay, P. R. (2002). *Main currents in western environmental thought.* Indiana University Press.

Howard, P. L. (2003). The major importance of "minor" resources: Women and plant biodiversity. *IIED Gatekeeper Series, 112.*

McGuire, S. (2015). FAO, IFAD, and WFP. The State of Food Insecurity in the World 2015: Meeting the 2015 International Hunger Targets: Taking Stock of Uneven Progress. Rome: FAO, 2015. *Advances in Nutrition*, *6*(5), 623–624.

Merchant, C. (1989). *Ecological revolutions: Nature, gender, and science in New England.* University of North Carolina Press.

Montanari, B., & Bergh, S. I. (2019). Why women's traditional knowledge matters in the production processes of natural product development: The case of the Green Morocco Plan. *Women's Studies International Forum*, *77*, 102275. Retrieved from: https://doi.org/10.1016/j.wsif.2019.102275

Morocco – Agricultural Sector. (n.d.). International Trade Administration | Trade.gov . Retrieved from: https://www.trade.gov/country-commercial-guides/morocco-agricultural-sector

Morocco. (2016). Troisième Communication Nationale du Maroc à la Convention-Cadre des Nations Unies sur les Changements Climatiques [Third National Communication of Morocco to the UNFCCC]. Ministère délégué chargé de l'Environnement. https://unfccc.int/documents/128109

Moyer, J. (2001). Why Kant and ecofeminism don't mix. *Hypatia*, *16*(3), 79–97. https://doi.org/10.1111/j.1527–2001.2001.tb00925.x

Mukherjee, U. P. (2010). *Postcolonial environments: Nature, culture and the contemporary Indian novel in English.* Palgrave MacMillan.

Pascon, P., & Ennaji, M. (1986). *Les paysans sans terre au Maroc.* Editions Toubkal.

Sachs, C. E. (2019). *Gendered fields: Rural women, agriculture, and environment.* Routledge, Taylor & Francis Group.

Salem, A. (2021, November 13). *The negative impacts of colonization on the local population: Evidence ...* Retrieved from: https://adres2022.sciencesconf.org/data/pages/Salem_JMP.pdf

Sandilands, C. (1997). Mother earth, the cyborg, and the queer: Ecofeminism and (more) questions of identity. *NWSA Journal*, *9*(3), 18–40.

Sheldon, K. (2013). Women and colonialism. *Oxford Bibliographies Online Datasets.* Retrieved from: https://doi.org/10.1093/obo/9780199846733-0067

Shiva, V. (1988). *Staying alive: Women, ecology and development.* Zed Books.

Shiva, V. (2016). *The violence of the Green Revolution: Third World agriculture, ecology, and politics (Culture of The Land)* (Reprint). University Press of Kentucky.

Stanley, A. (1981). Daughters of Isis, daughters of Demeter: When women sowed and reaped. *Women's Studies International Quarterly*, *4*(3), 289–304.

Varanasi, A. (2022, September 23). *How colonialism spawned and continues to exacerbate the climate crisis*. State of the Planet. Retrieved from: https://news.climate.columbia.edu/2022/09/21/how-colonialism-spawned-and-continues-to-exacerbate-the-climate-crisis/

Velde, P. van de, Stanley, V., & Stickler, M. M. (2020, October 14). *Invisible farmers: Why recognizing and supporting women farmers is key to food and nutrition security*. World Bank Blogs. Retrieved from: https://blogs.worldbank.org/developmenttalk/invisible-farmers-why-recognizing-and-supporting-women-farmers-key-food-an

Walsh, M. K., et al. (2020). *Climate indicators for agriculture*. USDA Technical Bulletin 1953.

Warren, K. J. (1990). The power and the promise of ecological feminism. *Environmental Ethics*, *12*(2), 125–146.

Warren, K. J., & Perlez, J. (2010). Inequalities plague African women. In Karen J. Warren (Ed.), *Ecofeminism: Women, culture, nature*. essay, Indiana University Press.

Warren, K., Erkal, N., & Curtin, D. (1997). Women's knowledge as expert knowledge: Indian women and ecodevelopment. In K. J. Warren (Ed.), *Ecofeminism: Women, culture, nature*. Essay, Indiana University Press.

Whitehead, A., & Tsikata, D. (2003). Policy discourses on women's land rights in sub-Saharan Africa: The implications of the re-turn to the customary. *Journal of Agrarian Change*, 3(1–2), 67–112. https://doi.org/10.1111/1471-0366.00051

Women, U. N., Khuvung, M., & House, U. N. (2020). Copyright© UN Women 2020. All rights reserved.

World Bank. (2022). *Agriculture, forestry, and fishing, value added (% of GDP) [Data set]*. World Development Indicators. https://data.worldbank.org/indicator/NV.AGR.TOTL.ZS

World Bank Group. (2018, March 6). *Breaking the "grass ceiling": Empowering women farmers*. World Bank. Retrieved from: https://www.worldbank.org/en/news/feature/2018/03/06/breaking-the-grass-ceiling-empowering-women-farm

15

Women and Land Property Within an Ecofeminist System

Redouane Bougtib

Ecofeminism entails an inextricable relationship between feminism and the radical ecology movement to facilitate justice for women and for nature at large (Hamad, 2013). Ecofeminism aims to establish new gender relations between humans and nature, and between men and women (Merchant, 1990). It emerged as an academic discourse (Carlassare, 1994) to spotlight women's efforts to subsist for themselves and their families in the face of patriarchy, capitalism, and environmental degradation (Murphy, 1997). It is therefore concerned with providing feminist responses regarding environmental destruction, as this destruction is perceived to perpetuate continuing types of oppression (Collard & Contrucci, 1989).

Ecofeminism is an intellectual and social movement like other movements that call for justice and equity for women and nature (Laplonge, 2016). This movement emerged in the 1970s and included environmental protection in parallel with women's oppression. Carolyn Merchant was the first to write about the relationship of exploitation of nature and women at the detriment of their existence and loyalty. Also, Françoise d'Eaubone was the first radical feminist to refer to ecofeminism in 1974 (Buckingham, 2015).

The term feminism (Lorber, 1997) holds a special meaning for users, based on history, gender, and positionality. Every individual can form a baseline definition of feminism. One could state it is concerned with the perceived inferior position of, and with women's encountered discriminations because of their gender. Feminism has diverse and multifaceted meanings. Another could argue that feminism calls for social, political, and economic change to minimize and overcome discrimination against women. It seems difficult to come up with a common ground of standards for feminism, depending on the ideology and theory used. Hence, feminism emerged after women started questioning their inferior status and demanding rehabilitation of their social position as agents of change.

However, today's capitalist world relies on technology and consumption for more interests at the micro and macro levels. This economic system has significantly altered people's perceptions regarding their living conditions and the adoption of competitive mainstream discourses, such as consumption-free and women's agency. This paved the way for new values to emerge, such as the norms of a modern capitalist world that go hand in hand with insatiable desires for extreme exploitation of natural resources. It encompasses every aspect of man's life, such as food, dress code, and work.

In this chapter, I examine the relationship between ecofeminism and Moroccan women in the Al Gharb region and in the Oulmes Mountains. These are two completely different cultures with varieties of languages. In the former region, women speak Moroccan Arabic, and some of them have been to school. In the latter, they speak a Middle Atlas variant of Tamazight and the majority have never been to school except for their young girls. In this territory, illiteracy and customary law (Peyron, 2018) are two factors that have allowed patriarchy to control these women's sustenance and mobility. In the Al Gharb region, advocacy associations and literacy programs enabled more women to call for their rights, such as the right to inherit a share of collective lands (Elaradi sullaliya). Such activism has inspired more women because other women in Mehdia were granted their land lots. They have become epitomes to others in Sidi Kacem province.

In the Al Gharb region, women broke free from patriarchal oppression and started calling for their right to cultivate their ancestral land. However, in the Oulmes Mountains, women are not allowed to call for their share of collective land because this community is deeply controlled by customary law, which is an indirect translation of a patriarchal thought. Males relate land to the sacred, which is not allowed to be granted to women, who might inherit it to their children or sell it. These two different cultures and territories yielded contradictory mentalities regarding land ownership and women's subordination as well. One commonality between the two regions is that males exploit nature and women for their advantage. Ecofeminism has connected males' exploitation of nature to women's oppression.

Females cultivate land to nourish their families and themselves and they do not intend to harm their environment at the detriment of their well-being.

Unlike males, who over-exploit nature by all means, such as deforestation, drying up of wells, and degradation of soil fertility because of excessive use of chemical fertilizers. In this article, I emphasize women's role in protecting their land/environment through the use of natural fertilizers, wise water usage, and long-term food provisioning thanks to their knowledge of crops, land, and cattle needs.

Methodology

This article aims to discuss how the capitalistic world has led more women to become aware of their rights to land and contributions. I attempt to compare these two different regions in terms of patriarchal domination, women's awareness of inequality, and the impact of climate change on these women's living conditions. The purpose of this paper is to better understand how patriarchy and climate change alter these women's identities, whether they are married or unmarried. The focus on women as agents of change is based on a feminist-Marxist theory of social change (Engels, 1972). Social development is approached from a Marxian feminist perspective of existing inequalities within the household and the community. This theory redefines contradictory relationships on the basis of recognizing a woman's role in bringing change and development. I adopt a qualitative approach by conducting ethnographic fieldwork in Sidi Kacem and the Oulmes Mountains, either at participants' homes or the weekly market. The data were collected through archives, participant observation, and semi-structured interviews. All fifty participants were asked to grant permission to record the interview. All the interviews were in Moroccan Arabic and Middle Atlas variant of Tamazight, then transcribed and translated into English. Confidentiality was emphasized throughout contact with participants, and they were told they could withdraw their recordings at any time.

The first section of this article explores the impact of patriarchal practices and illiteracy on these women's agency in both regions. Next, I will examine how female feminist activism has helped these women call for emancipation of themselves and nature as well. Finally, the last section highlights awareness of the necessity of women's and girls' education regarding land property for more economic opportunities to achieve their objectives as independent women within an ecofeminist system.

Fieldwork

My fieldwork was conducted in the Oulmes Mountains (Ait Ikkou) between April 2021 and June 2022. I also spent 8 weeks in Sidi Kacem (Bir Taleb). Upon arrival, one cannot help but be taken in by the beauty of the oak and pine-covered mountain ranges.

The Oulmes Mountains: AitIkkou Region

The statistics collected from the Agriculture Chambre in Khemisset region show that mountains represent 60 percent of this territory, plateaus 15 percent and plains 25 percent. Private property land represents 95.8 percent, collective lands are 1.7 percent and private domain of the state is 2.4 percent.

The area has known a significant stretch of modern houses made up of cinder blocks instead of mud blocks. The invasion of modern houses with two floors is significantly noticed in some places when moving from the village center to the mountains. Locals explain that only wealthy outsiders who prefer to spend their weekends in nature can afford to build such a house with a large farm surrounding, regardless of the price. This territory is also known for extremely cold winters where the temperature can drop to 7° or 0° Celsius, and in summer it can reach 35° Celsius, according to the chamber mentioned above.

The 2014 Census provided more information about the demography of this territory (Ait Ikkou). The population total is 8,865 (49.8 percent males, 50.24 percent females). The average size per family is 4.4 percent individuals. The school-age categories are as follows: 343 children between 4 and 5 years old, 986 pupils between 6 and 11 years old, and 492 pupils between 12 and 14 years old. The illiteracy rate of inhabitants more than 10 years old is 60 percent (47.5 percent males and 72.3 percent females). The average age of marriage is 31 years old for males and 23.8 years old for females, where 61.6 percent of the population older than 15 years old are married and 28.6 percent are unmarried, 2.2 percent are divorcees, and 7.6 percent are widows.

Ninety-nine percent of the population speaks a Middle Atlas variant of *Tamazight*, while those who speak *Tashelhit* make up 2.6 percent and *Tarifit* are 0.1 percent, 79 percent of this community uses Moroccan Arabic on a daily basis with non-natives.

Sidi Kacem: Bir Taleb Region

The statistics collected from the 2014 Census provide more information regarding this territory. The area has a Mediterranean climate between October and April, where humidity does not exceed the average temperature by 16 degrees. The second half from May to September experiences drought and intense heat, with an average temperature reaching 24.7 degrees Celsius. Most houses are modern due to the income locals earn from agricultural activities. The statistics provided by HCP show various types of land property in this territory as follows: Private property land represents 65 percent, Collective lands are 12 percent, Guishare 21 percent, and Houbousare 2 percent.

According to the 2014 Census, the population total of this territory is 11,370 (5,765 males, 5,605 females). The average size per family is 5.6 individuals. 96.5 percent of school-age children are between the ages of 7 and 12 years. (97.7 percent males and 95.6 percent females). The illiteracy rate of inhabitants is 42.7 percent (29.8 percent males and 56.1 percent females). The average age of marriage is 31.2 years old for males and 25.2 years old for females. 99.9 percent of the population speaks Moroccan Arabic, while those who speak Tashelhit are 0.2 percent, Tamazight are 0 percent, and Tarifit are 0 percent. The languages learned by the population are as follows: Standard Arabic is 42.2 percent, Standard Arabic and French are 51.8 percent, Standard Arabic and French and English are 5.2 percent, and others are 0.8 percent.

Description of Results

In both field sites, I interviewed males and females, asking them the same questions. The only difference was the language used. For the sake of clarification, (F) refers to the female group, while (M) refers to the male group. The interviews took place in two different geographical, cultural, and linguistic field sites. I conducted interviews in various locations within both territories, either at participants' homes or in the weekly market. Mastering the Middle Atlas dialect allowed me to communicate easily with locals in the Oulmes Mountains. Similarly, my proficiency in Moroccan Arabic facilitated communication with participants. I asked specific questions regarding women's inclusion in land possession, primarily collective lands and benefits. All participants from both field sites expressed their willingness to participate and share their views (see Table 15.1).

While interviewing every participant, I kept in mind the importance of education in opening up horizons for more socio-economic and ecologic projects that will benefit locals of all ages. After spending 14 months in the Oulmes Mountains, I found that locals value education, and they are not against educating their young girls. However, HCP (Haut Commissariat de Planification) admits that only 0.7 percent are able to go to university due to economic reasons and job scarcity. Accordingly, old and young generations share a common point

Table 15.1 Data of Oulmes Mountains and Bir Taleb

Sex	The Oulmes Mountains	%	Bir Taleb	%
F	1. Do you have your own land inherited from your parents?	40	1. Do you have your own land inherited from your parents?	90
F	2 Do your brothers agree to inherit land?	30	2. Do your brothers agree to inherit land?	100
M	3. Will you give your sisters their land share?	95	3. Will you give your sisters their land share?	100
F	4. Do you have access to collective land?	0	4. Do you have access to collective land?	100
F	5. We have no idea about collective lands	99.99	5. We know we have right to collective lands	100
M	6. Collective lands are governed by customary laws and the tribe notables	90	6. Women and girls have their right to collective lands according to 62-17 law	100
F	7. I herd the sheep and work in the household and implant some vegetables on our land	30	7. I know everything about farming, taking care of sick sheep or cow such as injection. I learnt this when the vet comes to consult the sheep and he showed me how to do it. My land is my life and my identity. I cannot live away from land.	100
F	8. I would like to move to city because living is getting hard because of water scarcity and unemployment.	90	8. I will not leave my land and my place because I was born here and I will stay here.	100

regarding education, which is the key to bringing success in one's life and to the rest of the community.

I also asked a question regarding women's right to collective lands and their socio-economic contributions. 100 percent of the men I talked to in both field sites stated they value the work women do either indoors or outdoors. Along with this, in the Oulmes Mountains, since collective lands represent only 1.74 percent and the areas are mountainous, the women interviewed confirmed they know nothing about this type of land ownership, and they will not confront their siblings, relatives, and tribe. They know every year's harvest is not as the previous year's anymore because of climate change and continuous years of drought.

At the other extreme, in Bir Taleb, terrains are vast and waters are abundant. Locals, after the visit of Ivanka Trump to launch the Melkization process of collective lands to Sullaliyates, create a deep social change, especially the cut with patriarchal practices that deprive women of access to collective lands as right holders (Berriane, 2021; Dupret & Buskens, 2021). Collective lands have been regulated by elder councils promulgating local laws that legitimize women's exclusion from collective lands. In Bir Taleb, all the women and men interviewed were happy to have legal access to their share of collective lands as their brothers. The questions I asked in every interview were the result of many previous discussions and meetings with right holders, feminism activists, and pre-schooling teachers, attending a Sullaliyates conference in Rabat on January 9, 2022. This allowed me to wear gender lenses and look at the issue of women's oppression and environmental exploitation by men and capitalism that attempt to exclude rather than include. Large areas were deforested to plant avocados, strawberries, or tangerines at the cost of polluted groundwater and women's deprivation of their legal right to collective lands (Hamdcach, 2020).

The statistics provided in the interviews above show there is a significant difference in terms of geography, culture, and benefits. In the Oulmes Mountains, all women are unaware of their right to collective lands and benefits as their male counterparts. They think this is only for men and the council of elders is in charge of regulating collective lands. These women know only private lands inherited from their parents as '*melk*'. Another significant factor that prevents them from recognizing is illiteracy and customary law. There are some active women's associations, but they are interested in making couscous and traditional carpets or selling chicken and medicinal plants in the local weekly market.

In Bir Taleb, all the women I interviewed have been to school, and this helped them to become active in local associations and benefit from government initiatives. A 46-year-old woman turned her house into a school in order to teach the other women in the village to read and write, and she appreciates this job as a teacher fighting illiteracy. The other woman is 30 years old, and she is also a teacher of preliminary education. However, despite the advantages of this social change in these women's lives, the environment is suffering more than before because of the over-exploitation of ground waters and the use of insecticides. This will have very destructive effects on the fertility of land and human life as well.

Discussion of Results

The existence of many forms of social identity exists to reflect the different ways in which people connect to other social categories. Intersectionality is a key concept in feminist theory, used to conceptualize the relation between systems of oppression which construct our multiple identities, our social location, power, and privilege. The aim is to clarify the need for intersectionality in our lives as a logical extension of critical feminist approaches. Hence, intersectionality has provided theoretical and political implications to uncover strategies of power, forms of oppression, and experiences of white middle-class women vs. women of low class.

Intersectionality is a framework to conceptualize people or social problems as impacted by a number of advantages and discriminations. It takes into account the overlapping of people's experiences and identities to comprehend the complexity of prejudices they encounter. The term intersectionality was coined by Crenshaw (2018) in her paper "Demarginalizing the Intersection of Race and Sex: A Black Feminist Critique of Antidiscrimination Doctrine, Feminist Theory and Antiracist Politics." The theory emerged after the white, middle-class feminist movement. Many black women found it quite difficult to identify with the white mainstream feminist movement. They were excluded from leadership positions and experienced sexism. The intersectional experience encouraged black women to call for a feminist practice that centralizes their lived experiences. Moreover, intersectionality is a feminist theory that starts from the premise that people have multiple identities according to their social history, relation, and operation of power structures. The context plays an important role in defining the features of one's identity. Experiences of oppression and privilege differ from one person to another, one group to another; this can reflect the layered identities a person can have simultaneously. For example, a woman can be a respected doctor and suffer from domestic violence at home. This paradox shows the proportionality of multiple lived identities we can have internalized.

Intersectional analysis exposes the different types of discrimination and disadvantages of a combination of identities. This analysis addresses the ways in which patriarchy, racism, class oppression, and other systems of oppression structure women's positions. The experience of an illiterate Black woman is quite different from that of a white educated woman. The two categories have different experiences that are dictated by various structures of power, systems of oppression, and privileges. Context and social relationships help to show and understand the continuity of white women's superiority over the black/low paid. However, intersectional analysis aims to overcome discrimination and put all people in the same rank to enjoy the same rights equally. Those that have multiple identities are pushed into the margins while others benefit from the privileges given by their social status. Intersectional analysis helps to understand and have a value judgement on progress, laws, and policies to see their impact on women's multiple identities. Intersectionality is therefore a tool to build a global culture of human rights from the grassroots to the global level.

Intersectionality means the process by which we think about identity politics. It requires our joint efforts to think effectively of points of intersection and equal rights. It did come to reveal the meaningfulness of multiple identities within society as doctors, lawyers, workers, and mothers. It celebrates the diversity of human rights as a fundamental praxis to development and full social participation. Women from different social backgrounds should have at least the same advantages regardless of their race, ethnicity, and social status. They have to enjoy the global human rights, according to the UN conventions and resolutions that stressed the need to combat violence against women and poverty. Besides, the usefulness of intersectionality depends on how it is used to bring change. Each one of us should wear intersectional lenses to understand the necessity of ending oppression and discrimination towards women of color in particular.

Women's oppression can be explained by social structures of patriarchy that secure men's power over women. Many feminists called into question the commonplace of power that regards women as the passive. In this light of discerning the possession of power by men over women, Foucault (1995) explains that power is exercised rather than possessed. The Foucauldian analysis of power asserts that power is exercised outside the political realm to examine the workings of powers in women's everyday lives. Intersectionality theory, therefore, asserts that people enjoy disadvantaged by multiple oppression sources and discrimination such as: religion, class, gender identity, and race. Intersectionality has become a commonplace within feminist theory, that women's lives are constructed by multiple intersecting systems of oppression. Intersectionality is then offered as a theoretical and political framework to highlight and advance women's position regardless of their color or social status, through having access to global human rights and political active presence. Women's problems are then rooted in the convenient political representation to defend their rights. Politics and women's personal experiences are interrelated in a way that demands reconsidering power relationships.

In the 1970s, feminists challenged prevailing accounts of gender relations with the slogan "the personal is political." The personal is political expressing that women's problems are rooted in personal experiences, due to political and social settings. "The personal is political" became more popular with the publication of Carol Hanisch's essay Women's Liberation in 1970. Hanisch explained that women's personal experiences are more associated with power relationships, and the personal is inextricably political.

Hanisch argues that women's problems were political ones, as women's lives were already molded and defined by political and social interferences. The context in which women live imposed how their lives should be conducted, and they were not granted enough personal freedom to decide by themselves or discuss their own problems with each other. Society limited their existence by giving solutions to their problems unilaterally, under the umbrella of power relationships. However, women's problems could not be solved by personal solutions but essentially by changing legislations (via politics).

Accordingly, women and the environment have been oppressed by patriarchal practices. COP 27 in Egypt and other climate change conferences in the UN called for the need to keep up with environmental degradation and women's inequalities. Today, the need to wear gender lenses to deliver on issues regarding the environment and equality is increasingly becoming pivotal to achieving sustainable finance (Ben-Meir, 2019) objectives and to enhancing inclusive growth of financial and human capital. Ecofeminism has taken an intellectual position to address issues related to women and the environment, such as oppression and patriarchy (Buckingham, 2015).

In the Oulmes Mountains, women were absent from claiming their rights as holders. In this community, the customary law excludes local women's access to collective lands, and they do not recognize having the right to benefit every year.

Collective lands are types of lands shared collectively by right holders labeled Sullaliyan/Sullaliyates who are believed to hold a "patrilineal lineage relationship among members of an ethnic group with collective land ownership" (Eddouada, 2021). At the national level, after law 62.17, women have been able to claim their share as members belonging to a particular tribe. The state launched a province pilot in Kenitra, in Mehdia, to own Sullaliyates land parcels like their male counterparts. Moreover, this successful experience triggered other women across the country to create associations and have legal representation and status (Ait Mous and Berriane, 2016). ADFM, as one of the most active associations related to Sullaliyates activism, has given the highest importance to these women's exclusion from their right to collective lands (Hassan Rachik, 2016). In January 2022, I attended a conference in Rabat organized by ADFM, which encompassed women activist leaders from different regions of the country. They shared their experience as Sullaliyates with no right to collective lands because of the requirement of residence and stressed the value of an annual income even if it is only one symbolic dirham. In the end, the attendees and speakers viewed the customary law as the main cause behind their exclusion.

Conclusion

This article adds a significant contribution to understanding how women's possession of collective lands will create a new spirit of development in the area of Bir Taleb, unlike where the customary law still denies women's right to collective land in the Oulmes Mountains. The article discusses this practice as an inherited legacy passed down from one generation to another in the community under study. Living in this territory enabled me to better understand the reasons why natives identified with such a practice despite its negatives. The council of elders of this territory has never discussed a woman's right to collective lands or given them the advantage to exploit them as a peasant. They still identify with the customary law to justify these practices and women's deprivation and exploitation. However, in Bir Taleb, natives' cultural practices have known some transformations regarding a woman owning collective lands. The Family Code and civil society activism have helped in granting girls and women their right to education and agency; however, abandoning customary law, delaying marriage age, and prioritizing education would cause some societal disorders in girls'" lives, such as an inferior position of an unmarried girl and a personal failure of inability to get married. Accordingly, along with the liberal process and policies, the state has undertaken mainly the Melkization of collective lands to attract more large landowners to invest in rural peripheries, and this would create more socio-economic inequalities to pop up and the locals would feel left out.

References

Aït Mous, F., & Berriane, Y. (2016). Droit à la terre et lutte pour l'égalité au Maroc : Le mouvement des soulaliyates [Land Rights and the Struggle for Equality in Morocco: The Soulaliyates Movement].In H. Rachik (Ed.), Contester le droit : Communautés, familles et héritage au Maroc (pp. 87–173). La Croisée des Chemins.

Ben-Meir, Y. (2019). Empowering rural participation and partnerships in Morocco's sustainable development. *Journal of Global Initiatives: Policy, Pedagogy, Perspective, 14*(2).

Berriane, Y. (2021). "La part des femmes": La judiciarisation de l'accès des femmes aux terres collectives au Maroc ["Women's Share": The Judicialization of Women's Access to Collective Land in Morocco]. Cahiers d'études africaines, 242, 417–438. https://doi.org/10.4000/etudesafricaines.34374 journals.openedition.org

Buckingham, S. (2015). Ecofeminism. *International Encyclopaedia of the Social & Behavioural Sciences, 2015*, 845–850. Retrieved from: https://doi.org/10.1016/b978-0-08-097086-8.91020-1

Carlassare, E. (1994). Destabilizing the criticism of essentialism in ecofeminist discourse. *Capitalism Nature Socialism, 5*(3), 50–66. Retrieved from: https://doi.org/10.1080/10455759409358597

Collard, A., & Contrucci, J. (1989). *Rape of the Wild: Man's Violence against Animals and the Earth*. Indiana University Press.

Crenshaw, K. (2018). Demarginalizing the intersection of race and sex: A Black feminist critique of antidiscrimination doctrine, feminist theory, and antiracist politics [1989]. *Feminist Legal Theory*, 57–80.

Dupret, B., & Buskens, L. (Eds.). (2021). *State law and legal positivism: The global rise of a new paradigm*. Brill. ISBN 978-90-04-49865-5.

Eddouada, S. (2021). Land rights and women's rights in Morocco: Cooperation and contestation among rural and urban women activists. *History of the Present, 11*(1), 23–52.

Engels, F. (1972). *The origin of the family, private property and the state*. New York, NY: International Publishers. (Original work published 1884)

Foucault, M. (1995). *Discipline and punish: The birth of the prison*. Vintage.

Hamad, R. (2013). Ecofeminism. *Redress, 22*(2), 11–15.

Hamdcach, A. (2020). ةسارد ريرقت :بلاطمو عاضوأ :برغملاب تايلالسلا ءاسنلا ةيجولويسوس [Sullaliyates women in Morocco: Conditions and demands: A sociological study report].

Laplonge, D. (2016). Exploring the distance between ecofeminism and women in mining (WIM). *The Extractive Industries and Society*, *3*(3), 843–849.

Lorber, J. (1997). *The variety of feminisms and their contributions to gender equality*. BIS Verlag.

Merchant, C. (1990). *Ecofeminism and feminist theory*. In I. Diamond & G. F. Orenstein (Eds.), Reweaving the world: The emergence of ecofeminism (pp. 100–105). San Francisco: Sierra Club Books.

Murphy, P. D. (1997). Ecofeminism and postmodernism: Agency, transformation, and future possibilities. *NWSA Journal*, *9*(3), 41–59.

Rachik, H. (2016). *Contester le droit: Communautés, familles et héritage au Maroc*. La croisée des chemins.

16

The Two Cultures, the Third Culture, and the Environment in Ian McEwan's *Solar*

Nasreddine EL Guezar

It is globally believed that the role of academia is to make for an educated human being who can think critically and holistically about the most relevant issues of the day. Nonetheless, one notes an enormous embranchment of disciplines that has been increasing at an alarming rate. Academics, thus, have been facing challenges in their ability to thoroughly comprehend the vast and rapidly growing fields of knowledge nowadays. Over-specialization has facilitated the expansion of gaps in many domains of knowledge, and disciplinary examination has often rendered scholars and specialists unable to tackle issues outside their fields of expertise. As knowledge disciplines are kept unconnected, academia and academics have become inadequate at solving many problems of the day, including environmental ones. In other words, ongoing academic gulfs are a hindrance to the process of finding reliable solutions to the most vexing planetary problems of the day. Our shared fate and the future of our planet are a concern for all of us. Therefore, academics from various fields, including sciences, social sciences, arts, and humanities, need to collaborate more closely. To gain a better understanding of this issue, we consider how this subject is approached in fiction.

Concerns about ongoing conflicts within academia have driven the proliferation of innovative and artistic ways to address these issues. Through the medium of fiction, numerous writers have endeavored to creatively conceive plausible events and characters as a means of addressing the shortcomings in knowledge and narrow disciplinary perspectives that currently plague academia. Among other novels, Ian McEwan's *Solar* is a substantial piece of fiction to explore in this regard. The satirical novel abundantly depicts themes and characterization of academic divisions. Following him on his substantial mission to tackle climate change, the protagonist, Michael Beard is described as having a narrow-minded perspective and being resistant to interdisciplinary collaboration. The novel does not only portray academic divisions, but it also highlights some important possibilities to bridge gaps and lessen these divisions, particularly through characters that have been following a more holistic approach in their dealings with the environment. Both the divisions and opportunities for connection are worth addressing as they play a significant role in the pursuit of solutions to the environmental problem of climate change in the novel. To delve deeper, this paper explores the aspects of knowledge divisions and connections in *Solar* through the lens of the ecocritical approach.

Ecological literary criticism encourages the exploration of academic hostilities and their environmental implications in literary fiction. Among the specific questions that ecocritics and theorists ask is "How is science itself open for literary analysis?" (Glotfelty & Fromm, 1996, p. XIX). In addition, the pioneer ecocritic, Glen Love, states, "A nature-conscious, nature-validating literature and criticism offers a needed corrective in its regard—either implicit or stated—for the nonhuman" (2003, p. 26). Ecocriticism then revalues literary texts which are needed to explore a profound dimension to the current environmental questions. Academic divides are among these dimensions. Relevantly, ecocriticism is known to do the interdisciplinary work of making connections. It generally brings together ecology, which is essentially a science, with the study of literature and humanities. An important contribution of ecocriticism is, therefore, to bridge this gap. Consequently, bringing the analytical work of literary criticism to the ecological discussion not only connects disciplinary breaches and unfortunate gaps in knowledge but also presents us with new cultural and critical insights, reflecting positively on the planetary dilemmas. Since Ian McEwan's *Solar* portrays many knowledge gaps and opportunities for connection to the environment, it is useful to read the novel from an ecocritical perspective, particularly in light of what is referred to as the two cultures and the third culture.

It is pertinent to refer back to C. P. Snow's *The Two Cultures*, which has brought into prominence the dividing line between the two massive knowledge fields, science and the humanities, paving the way toward the bridge of the third culture. In the same vein, the interdisciplinary study of literary texts from an environmental lens goes in this direction. For that matter, in analyzing *Solar*, it looks at the novel in light of the two-three cultures' academic debates that C. P. Snow and subsequent scholars have raised. Their theoretical contributions make for useful lenses to examine the attitudes and the environmental implications of the characterization in the novel. Through the character Michael Beard, Ian McEwan's *Solar* portrays a world where science stands in clear opposition to others domains, mainly the humanities, and where it is unable to address environmental issues. On the other hand, Tom Aldous—Beard's opponent, seemingly—appears to set the example of the qualities that Michael Beard specifically needs but lacks. Nevertheless, Beard has been unwilling to work with Aldous and veered away from a potential complementary relationship. As a result, and because Michael Beard has been overly hostile to opportunities for connection, his research fails to make the promising effects on the environment that he has striven for. Knowledge gaps and indifference to connections are then considered principal causes for the failure of scientists in *Solar* to save the planet and humanity from an impending environmental issue.

The Two Cultures and the Third Culture: A Brief Overview

As a public intellectual, having one foot in science and another in the humanities, Charles Percy Snow has been, in a manner of speaking, eligible to talk about the knowledge gaps that dominate academia. Snow contends that between the literary intellectuals and scientists exists "a gulf of mutual incomprehension—sometimes . . . hostility and dislike, but most of all lack of understanding" (2012, p. 4). Accordingly, feelings of resentment rise between the two sides when they fall short of proper communication. In his book *A Second Look*, Snow elucidates that "Between these two groups—the scientists and the literary intellectuals—there is little communication and, instead of fellow feeling, something like hostility" (2012, p. 61). In so doing, Snow has helped in paving the way to what he terms to be the "third culture" to refer to possibilities that could end the incomprehension (2012, p. 70). By highlighting the divide, Snow has laid the foundations for subsequent attempts that aim at closing the gap between science and the humanities.

Part of the hostility between the two cultures results mostly from the lack of comprehension, and the rise of popular science mitigates that. In his book, *The Third Culture: Beyond the Scientific Revolution*, John Brockman gathers a few scientists in one place to express their insights on different complex topics in an accessible manner. This way, the scientists have made themselves comprehensible to the non-scientists. When science is explained in accessible terms, comprehension increases, and the hostility, which Snow has warned of, diminishes. Quite similarly, the American biologist and renowned naturalist, Edward O. Wilson, argues for taking the natural sciences as an umbrella for all other branches of learning. It is a unity where fields of social science, arts, ethics, and religion are drawn into an explained, fact-based, and rational science. By doing this, the linkage of natural science and the humanities happens without falling out to "subjective" thinking which, according to Wilson, only hinders the pursuit of truth (1999, p. 73). Thus, the key to unifying knowledge, to Wilson, lies in the "consilience" of all knowledge fields under the science banner. Consilience under the umbrella of natural science could be a practical proposal for uniting branches of knowledge in trust of rationality, reductionism, objectivism, and empiricism.

Despite its advantages, it is difficult to make up the minds of thinkers in the humanities about Wilson's consilience. Skipping theories that are not factually, empirically, or materially explained disregards much expertise in literature, history, culture, philosophy, and the rest of the disciplines in the humanities. The version of unity Wilson calls for does not appreciate the element of difference and primarily cares for creating one unified knowledge that turns every discipline into an exact science. George Levine's *One Culture: Essays in Science and Literature* composed of twelve essays that bring together historians, literary critics, and philosophers of science to explore the relationship between science and literature. These writers apply significant analyses and interpretations from literature and the social sciences to science itself. As a result, the articles in the collection demonstrate how science and the humanities are siblings, belonging to "one culture" rather than two. Science and literature are incompatible due to differences in the nature of language discourse and definition, and Levine does not deny these differences. Rather, he, along with other writers in the volume, attempts "to understand those differences and perhaps discover ways that the enterprises can be seen as mutual" (1987, p. 4). Because Levine's *One Culture* presents compelling evidence that science is as socially and culturally embedded as any piece of literature, "Science is reflected in literature. And the tools of literary criticism become instruments in the understanding of scientific discourse" (1987, p. 22).

Levine might have been influenced by the English critic F. R. Leavis, who harshly reacted to Snow's concept of the two cultures. Leavis's "ad hominem" critique seems to bring misanthropes toward him; however, it could also be useful. Most who read Leavis's response for the first time may deduce that he has been insolent and academically bad-mannered. However, what Leavis has meant to disrespect is someone preaching with unquestionable confidence on a topic he might be unqualified to talk about. He encouraged the public to be more critical in their assessment of those who make grand statements and not to accept them without question. Leavis reminds us that literary critics have in "their bones" the ability to explore deeper meanings and complex dimensions that laymen and scientists do not possess. By questioning the foundation upon which Snow was established as a "sage and intellect" and a representative of both groups, Leavis alerts us to the misleading authoritative tone that manipulates the public to attain unchallenged acceptance. Leavis's position, as a result, revalues the role of literature and literary critics. As Leavis notes, "The 'literary world' is its own public—the only one it, normally, is conscious of. It hates the suggestion that there might be another, one to be feared: a real educated public that doesn't take the 'literary world' seriously" (2013, p. 85). Leavis, in so doing, asserts that the literary world needs to be valued instead of being menaced and undermined.

The role of literature has become important to the science-humanities environmental discussion. Ecocriticism, along with other emerging interdisciplinary and multidisciplinary fields in the environmental humanities, has been meant to mitigate knowledge gaps as their effects are hindering the process of addressing larger environmental challenges. Concomitant with this, literature and literary language specifically contribute to this hoped-for collaboration since it helps "us, scientists and laypeople alike, to appreciate the *meaning* of our environmental quandaries" (Slovic, 2008, p. 147). "Environmental literature shows us how to 'see' the places where we live, revaluing what we've overlooked or ignored, and entering into more reciprocal relationships of care" (Hubbell & Ryan, 2021, p. 183). For that matter, it is of paramount importance to explore nonscience and literary material that enriches the environmental debate. Relying on the theory of two-three cultures, the subsequent chapter attempts to examine *Solar* and the environmental implications of the existent gaps and connections in the novel as portrayed through two essential characters, Michael Beard and Tom Aldous.

Analyzing the Two Cultures and the Third Culture in *Solar*

The story revolves around Michael Beard, a troubled scientist who tries to solve climate change. Beard is married for the fifth time and cannot confront his current wife, Patrice, who cheats on him. His scientific work and research are declining, unlike that of the excited young researcher, Tom Aldous, who has more ambitious ideas. Beard goes with a group of artists to observe the development of climate change in the Arctic. When he gets back, he finds Aldous having an affair with his wife. He confronts him and accidentally kills him. Beard, afterward, takes Aldous' notes and ideas, pertaining to solar energy and builds on them to revive his scientific career. As the novel approaches its end, Beard's attempt to find a solution to the world's environmental change fails.

Both the two cultures' divide and the third culture's bridge are abundantly revealed in the novel by Ian McEwan. *Solar* depicts how the environmental issue of climate change is approached by scientists. In doing so, the novel seems to accentuate numerous instances of animosity within the academic sphere. A clear resistance against many fruitful opportunities for collaboration is essentially highlighted through the character of Michael Beard, who shows unreasonable support for new or different opinions and plans of action. Unlike Beard, Tom Aldous is seen leaning to the bridge side. He is more likely to see connections and welcome them. The relationship between the two is characterized by resistance and integration, which is meant to communicate hindrances that stand in scientists' way of approaching substantial environmental challenges.

Michael Beard and the Two Cultures

Solar's Michael Beard has much to do with the divide between science and the humanities. It can be stated that Beard is himself a representation of the hostile gap between the two cultures and, in turn, a contributor to it. Right from Beard's past, one notices the roots of the divide he exemplifies. The characteristics of the two cultures' divide have always been inextricable from Beard. Even when he was still a student at Oxford High School, Beard would avoid students of the humanities and the liberal arts as he could not comprehend them: "and he also avoided the artsy sort of people—they intimidated him with literary references he did not understand" (McEwan, 2010, p. 227). This gap has been taking place years before Beard becomes a renowned scientist. This separation between the two cultures is then old and rooted as characterized through Michael Beard.

The incomprehension occurring among the scientists reveals further academic gaps through Michael Beard. The latter is unable to understand the language of the young undergraduates, referred to as "ponytails" whom he meets at the research center. "The ponytails spoke at speed, on a constant rising interrogative note, which caused an obscure muscle to tighten in the back of Beard's throat as he listened. They failed to enunciate their words, going only far with a thought, until one of the others muttered, 'Right!'" (p. 24). The undergraduates seem comfortable talking about some physics that is unfamiliar to Beard: "Some of the physics that they took for granted was unfamiliar to him" (p. 24). Beard could not comprehend the words articulated by the young undergraduates and has been unable to cope with them. Incomprehension, here, contributes to widening gaps in understanding, adding to the hostility and tension among scientists. As revealed, the incomprehension is not happening between science and the humanities, but one that happens among the scientists themselves, between old scientists and emergent ones.

The incomprehension in the novel aggravates even more with the coming of the postdoc, Thomas Aldous. The latter explains to Beard what he believes to be the solution to climate change: "[Solar energy]'s all out there, waiting for us to understand how to use it, and when we do, we'll be amazed we ever thought of burning coal and oil and the like" (p. 29). Beard, on the other hand, is unmoved and unwelcoming to Aldous' proposed solution. The division is augmented as we note further indifference occurring between scientists. Beard "hoped to avoid running to Aldous alone, for the young man was always trying to convince him of photovoltaics, or his quantum explanation of photovoltaics, or to oppress him generally with friendliness and enthusiasm" (p. 32). While Aldous brings novel ideas, Beard tends to avoid him and evade what would collide with his own version of solutions. This implies a lack of discussion and consideration for one another's thoughts among scientists in the novel. This adds more to the incomprehension and contributes to more gaps between Michael Beard and other scientists.

Through contributing to the divide, Beard renders himself inadequate in the face of the scientific matters he deals with. As depicted in the novel, the protagonist never grows academically, it keeps falling. As a womanizer and a "cuckold," Beard becomes obsessed with his wife whom he has been taking for granted. Only when she starts betraying him that he becomes occupied by a new love for his unfaithful wife. "What impressed him was his ability to think of nothing else. When he was reading a book, when he was giving a talk, he was really thinking of her" (p. 9). While he is acknowledged as a renowned scientist, Beard struggles with personal problems and is prone to live through unsettled marriages. In addition, he does not strive to understand and contemplate what can be done. "One thing was certain: two decades had passed since he had last sat down in silence and solitude for hours on end, pencil and pad in hand, to do some thinking, to have an original hypothesis, play with it, pursue it, tease it into life" (p. 16). Rarely do we spot Beard sitting alone at moments of reflection and contemplation over his attitudes and actions. Beard is then considered emotionally lacking. It is suggested here that Beard's inability to grow, on a personal level, springs from his inadequacy as a scientist who remains unreasonably determined to widen academic chasms.

Michael Beard has been inadequate in his inability to grow both personally and academically. He is generally obsessed with facts and rational thinking, which has led him to marginalize other modes of thinking and reflection. Michael Beard is depicted as a hostile scientist to academic groups from both science and the humanities. He is often immersed in a rooted incomprehension and rarely opens up to possibilities that may lessen the divide. This explains his marital failures and stumbling scientific career. It can be seen that the characterization of Michael Beard illustrates the two-cultures divide that C. P. Snow and other writers have disapproved of. Beard's attitudes, therefore, clearly contribute to the science-humanities breach and impede efforts to deal with the environmental challenges in a more thorough manner. McEwan's *Solar* then characterizes Michael Beard as a critique of academic divides and implicitly communicates the dangers of the two cultures' divide.

Tom Aldous and the Third Culture

Unlike Michael Beard, Tom Aldous demonstrates the bridge of the third culture. The characterization of Tom Aldous suggests he is more likely to connect gaps and mitigate the world's environmental problems than Beard. Aldous has had close contact with non-scientific fields. His sheer will to open up to broader possibilities, as a consequence, has made Aldous better equipped to treat environmental challenges. Among the suggested reasons for Aldous' ability to see possibilities of academic connections stems from his educational background that has been influenced by the insights of the humanities. Aldous' father was "a formidable history teacher" (p. 42). Although the relationship between Aldous and his father is not clearly revealed in the novel, it is suggested that roots of the humanities have been planted in Aldous from an early age.

Aldous' contact and familiarity with the humanities push him to call for more consideration toward non-scientific fields. "There were novels Aldous wanted him to read—novels!" (p. 33). Aldous demonstrates the existence of a will to bridge the gap between scientists and thinkers of the humanities. A scientist sitting to read novels is probably one of the exemplary forms of academic bridging. Aldous pushes in this direction after having experienced benefits through readings of the literary compositions. Nonetheless, it is not only novels that influence Aldous but also other art forms. Due to this, Aldous agrees that scientists need to look at the variety of perspectives that are plentiful in the humanities and liberal arts.

> There were novels Aldous wanted him to read—novels!—and developments in contemporary music he thought Beard should be aware of, and movies that were of particular relevance, documentaries about climate change which Aldous had seen at least twice but would happily see again if there was a chance of making the Chief sit through them too. (p. 33)

The mere act of a scientist watching a fictional movie or an artistic documentary about climate change may contribute to lessening the gap. Aldous, in effect, is noted to be aware of things that other scientists like Beard and the chief, Braby, are not. Aldous knows much about useful developments in music and has gathered much from his encounters with environmental movies and documentaries. This means that the knowledge and creativity Aldous has gathered come as inspirations from his contact with the various fields of the humanities and arts.

Aldous' openness to non-scientific contributions places him as a third-culture bridge. Filled with ideas he incorporates from broader knowledge fields and a clear willingness to connect academic gaps; Aldous believes that climate change and energy problems will end through benefiting from the solar energy we have everywhere. Aldous explains his idea in a thought-provoking analogy:

> There's a guy in a forest in the rain and he's dying of thirst. He has an axe and he starts cutting down the trees to drink the sap. A mouthful in each tree. All around him is a wasteland, no wildlife, and he knows that thanks to him the forest is disappearing fast. So why doesn't he just open his mouth and drink the rain? Because he's brilliant at chopping down trees, he's always done things this way, and he thinks that people who advocate rain-drinking are weird. That rain is our sunlight. (p. 31)

Trees here refer to commonly exploited natural resources that we grow dependent on, and we have become blind to the abundant "rain" falling all over us, the sunlight. Seen in light of the focus of this paper, the rain can refer to the possibilities of linking knowledge fields, and the division may lay in our inability to see that. That rain is the abundance that can generate beneficial and thriving connections between separate knowledge fields. Relevantly, this reflects positively on the environmental awareness of Aldous. While Beard is locked in his own ideas and blind to the evident possibilities, Aldous knows the planet is in danger and is more aware of what is at stake. Aldous utters to Beard how the idea of the micro-wind power is hopeless and can be replaced with the promising solar energy, "micro-wind is not going to solve the problem, Professor. The wind doesn't blow hard enough in most towns. We need a new energy source for the whole civilization" (p. 39). Michael Beard, however, cannot see possibilities existing in thoughts expressed by Aldous as long as he is restricted by the rigid divide between the two cultures. All in all, Aldous possesses more sense of the environment and the future threats than Beard due to his openness to the possibilities of the third culture.

Solar brings up the character of Tom Aldous to illustrate the outcome resulting from positive exposure to different backgrounds of knowledge. "Aldous had a mind that was designed, through the medium of a Norfolk accent, to offer tireless advice, make recommendations, urge changes, or express enthusiasm for some journey or holiday or book or vitamin, which itself was a form of exhortation" (p. 33). Aldous is filled with an enthusiastic will for exhorting influence and change. The suggested idea here is that the world needs more scientists like Aldous, scientists who exemplify bridges in place of gaps. The world needs more scientists who are developing a sense of complete awareness of what is at stake, who can spot things that go unnoticed, and people who are, in effect, able to approach problems and crises more holistically. In fact, it is through Aldous that we learn what the missing ingredient for solving world environmental challenges is in the story. Aldous seems to understand the importance of the humanities. Whether he is consciously or unconsciously advocating for the third culture connections, what matters is that Aldous has come to understand the usefulness of insights in the humanities to science; otherwise, he would not have shared his ideas with Beard nor would have urged the chief, Braby, to sit through the literary and artistic productions. Aldous contributes to the third culture, so to say. It is through Aldous that scientists, as it seems, are invited to pay attention to existing possibilities for connecting gaps and better approaches to the world's environmental challenges.

The Two-Third Cultures: Resistance and Integration

It can be said that the relationship between the two cultures' divide and the third culture bridge in *Solar* is characterized by resistance and integration. The two cultures tend to resist the third culture exactly as Michael Beard does toward Tom Aldous in the novel. This resistance is one of the major causes that lead Beard astray from finding the anticipated solution to climate change. Nevertheless, there are important signs of integration taking place between science and the humanities as well.

The resistant relationship between the two cultures and the third culture is revealed in many instances in the novel, especially at the time when Beard is the only scientist among artists. Beard goes on a journey to observe the development of climate change in the North Pole and finds himself the only scientist among a group of climate change artists. Beard tries to avoid the artists, contributing, in this way, to more gaps. "Beard, squashed in a corner . . ., hostile to groups, feigned a smile. He did not like to be part of a group, but he did not want the group to know" (p. 72). Beard resists any chance to connect with the artists. He even shows little caring "for art or climate change, and even less for art about climate change, he kept his thoughts to himself and was affable" (p. 84). Michael Beard's attitude expresses resistance toward the possibility of collaboration and holds a sceptical view toward art and any association with it.

The confrontation that Beard has with Tom Aldous reveals more of the two cultures' resistance. Aldous gives an inspirational speech to collaborate for the sake of the planet, yet Beard rejects his plea. Aldous implores, "we could skip all anger and heartache, we could be efficient about this, and we could be friends" (p. 99). He also adds, "It is too important to let go, it's our future, the whole world's future that's at stake, and that's why we can't afford to be enemies" (p. 102). The confrontation highlights how the resistance is so rigid and hard to exterminate. Beard is still immersed in his private problems that he disregards the possibilities existing to solve bigger problems. Aldous has always been trying to convince Beard to consider his ideas and build on his scientific establishments to end climate change and energy problems. Aldous, nonetheless, fails in face of two cultures that are profoundly disunited. Beard misses out on the depth of seeing possibilities of the third culture. Perhaps, if Beard had let go of some of his unbending resistance, he would have been able to collaborate for both his sake and the planet.

Though it is notably resisted by Beard, hints of the third culture bridge are fairly pervasive in the novel. When Beard sits in a committee in the seminar room at Imperial College, he is interested in listening to what a professor in science studies, Nancy Temple, says. Temple speaks on how the gene can be socially constructed and how it "can only come to us mediated by culture" (p. 152). Science studies are one form of the third culture as it tries to benefit from modes of understanding existing in other fields including the humanities. While it is a novel and resisted form, it greatly reveals how the third culture is happening.

Against his inclination to resist the bridge of the third culture, Michael Beard is slightly integrating into it. In his journey to the North Pole, Beard has been more open to the third culture. Even though he is typically not a fan of being among artists, he eventually indulges in conversations with people of the humanities and arts. Beard's journey to the North Pole with the group of artists illustrates how scientists and artists gather in one place to talk about world issues. Though he is usually hostile toward the artists, Beard has been somewhat destined to cope with the very third culture he is resistant to. When the sculptor, one of the artists, asks Beard questions like: "was the strange reality described by quantum mechanics a description of the actual world, or was it simply a system that happened to work?" (p. 75), Beard compliments the sculptor on his questions and answers them easily. The dialogue Beard has with the sculptor demonstrates a typical example of the communication bridge taking place. Beard is thus found merging into the third culture, and this is further illustrated in dialogues he has with the artists.

Perhaps Beard reaches his closest stage of integration with the humanities when he engages in reading John Milton. Beard finds special interest in the English major student, Maisie (his first wife), and to get close to her, he reads about her favorite poet, Milton. He reads many of his poems and essays until Maisie wonders "what a scientist was doing knowing so much about Milton" (p. 231). Beard "pretended . . . All knowledge interested him, he said; the demarcations between subjects were mere conveniences, or historical accidents, or the inertia of tradition" (pp. 231–2). Beard's hostility diminishes when he familiarizes himself with topics in the humanities. There is an illustration here that might suggest Beard is crucially benefiting from the humanities. Reading Milton has helped Beard understand certain concepts in his field.

> that very week [of reading Milton] . . . he studied the Ricci scalar and finally understood its use in general relativity . . . the theory was no longer an abstraction; it was sensual, he could feel the way the seamless fabric of space-

time might be wrapped by matter, and how this fabric influenced the movement of objects, how gravity was conjured by its curvature. (p. 233)

Reading poems by Milton has helped the student of science to grasp matters pertaining to science that otherwise would take longer to comprehend. This example and others within the novel clarify that the gap can be diminished.

Beard is not fully integrated into the third culture. Although Beard reaches a stage whereby he is building more communication with people of the humanities and liberal arts, his version of the third culture is found corrupt and lacking integrity. Scientists can indulge in philistinism—being uninformed of other knowledge areas—and disguise that through an environmental discourse just like Beard has done. "For humanity en masse, greed trumps virtue. So, we have to welcome into our solutions the ordinary compulsions of self-interest, and also celebrate novelty, the thrill of invention, the pleasure of ingenuity and cooperation, the satisfaction of profit" (p. 173). This is only one statement among many others throughout the novel showing how materialist a scientist's ends can be. It is suggested then that having a corrupted integration has been one of the causes that failed Beard's attempts to solve climate change.

Beard's project is derailed when plagiarism is discovered; most notably, it is discovered by a man in the humanities. Jeremy Mellon, a professor in urban and folklore studies, is listening when Beard talks about the promising project of solar energy. He could easily discern the absence of originality in Beard's talk. As a result, Beard is exposed in front of the attendees as a plagiarist scientist when he touches on something closely related to Mellon's field: "It is much studied in my field" (p. 182). One interpretation here could mean that any attempt to solve environmental problems does not work when it is not constituted on a morally solid foundation. It is suggested, therefore, that scientists like Beard might always need thinkers in the humanities, for the latter can determine the moral and humanistic validity of science. Hence, for solving environmental changes, scientists need to collaborate with modes of knowledge and disciplines beyond theirs.

In short, the two cultures' divide is increasingly shrinking even though it is faced with hostile resistance. Beard in the novel represents both an inflexible two-culture divide and a stiff resistance to the third culture. The failure of Beard means the disappointment of the divide. Perhaps, if Aldous were in place of Beard, he would have been able to function more properly and would have gotten closer to solving climate change. The story of Beard suggests that one cannot deal reliably with complex environmental challenges without first attending to knowledge gaps. What is needed is that the two cultures open fully and truly to the possibilities offered within the bridge of the third culture. Until we address the disparities in academia, it may remain difficult for scientists to effectively address environmental changes.

Conclusion

An ecocritical reading of McEwan's *Solar*, particularly through the lens of a two-three culture, demonstrates, through the character Michael Beard, hostile divisions among knowledge fields. Tom Aldous, on the other hand, serves as an example of the unconsidered bridge. The connection between scientists in the novel is notably characterized by resistance. This renders their scientific attempts inadequate to solve the world's environmental issues. Therefore, *Solar* suggests that scientists cannot properly handle environmental challenges without first ending the gap between science and the humanities. Resisting disciplinary connections deepens unfortunate knowledge gaps which in turn are seen to contribute to the failure of scientists in saving the planet and humanity from impending environmental catastrophes.

The academic and planetary concerns *Solar* brings up are linked to what many other researchers advocate. In their article, "Mapping Common Ground," Bergthaller et al. elucidate that "The ecological crisis is not only a crisis of the physical environment but also a crisis of the cultural and social environment" (2014, p. 262). Sorlin, in the same vein, contends that "in a world where cultural values, political and religious ideas, and deep-seated human behaviors still rule the way people lead their lives, produce, and consume, the idea of environmentally relevant knowledge must change." (2012, p. 788). That is why, scholars in the humanities advocate for "a deeper understanding of human beings and their varied relationships with the world in which they have evolved" (Bergthaller et al. 2014, p. 268). These gaps and opportunities for connections have facilitated the emergence of domains under the banner of what is often referred to as the environmental humanities. Scholars in this interdisciplinary field see the nature-culture divide as the root of the crisis and such a divide is compounded by academic divisions (Rose & Robin, 2004).

In a nutshell, efforts to mitigate environmental problems may continue to fall short as long as the gaps among knowledge fields are not fully mended. Narratives, such as *Solar*, exemplify a substantial need for considering these gaps and illustrate the possible risks of ignoring them for too long. As McEwan's *Solar* embodies within its storyworld, the scientist fails in his scientific mission due to an unfortunate resistance to opportunities for academic collaboration. The novel calls for disciplinary connections that could breed a more reliable understanding of environmental concerns. The looming environmental disasters facing Earth are dire and should bring together experts from all fields to work toward better explanations and solutions. Ultimately, knowledge has always been the key to guiding academics, politicians, policymakers, engineers, writers, and artists in addressing the challenges humanity faces. Therefore, a more comprehensive understanding of the issues at hand would serve humanity and help it move toward a more sustainable future for all.

References

Bergthaller, H., Emmett, R., Johns-Putra, A., Kneitz, A., Lidström, S., McCorristine, S., Pérez Ramos, I., Phillips, D., Rigby, K., & Robin, L. (2014). Mapping common ground: Ecocriticism, environmental history, and the environmental humanities. *Environmental Humanities, 1*, 261–276. https://doi.org/10.1215/22011919-3615505

Brockman, J. (1996). *The third culture*. Simon and Schuster.

Glotfelty, C., & Fromm, H. (1996). *The ecocriticism reader: Landmarks in literary ecology*. University of Georgia Press.

Hubbell, J. A., & Ryan, J. C. (2021). *Introduction to the environmental humanities*. Routledge.

Love, G. A. (2003). *Practical ecocriticism: Literature, biology, and the environment*. University of Virginia Press.

McEwan, I. (2010). *Solar*. Anchor Books.

Rose, D. B., & Robin, L. (2004). The ecological humanities in action: An invitation. *AHR – Australian Humanities Review*. Retrieved from: http://australianhumanitiesreview.org/2004/04/01/the-ecological-humanities-in-action-an-invitation/

Slovic, S. (2008). *Going away to think: Engagement, retreat, and ecocritical responsibility*. Basque Studies Program/322.

Snow, C. P. (2012). *The two cultures*. Cambridge University Press.

Sörlin, S. (2012). Environmental humanities: Why should biologists interested in the environment take the humanities seriously? *BioScience*, *9*, 788–789.

Wilson, E. O. (1999). *Consilience: The unity of knowledge*. Vintage.

17

Ecotherapy and Communion with Nature in Farid Alansari's *The Return of the Cavaliers*

An Ecotheological Perspective

Abdelaziz Dqaqi

The religious discourse has always been indicted for being an anti-environmentalist one. Different scholars have charged Abrahamic religions for being anthropocentric by providing theological grounds to abuse natural resources (White, 1967; Toynbee, 1972). These claims provided a major boost to a nascent ecological approach, namely Ecotheology or religious environmentalism. In this regard, scholars like Seyyed Hossein Nasr, Mary Evelyn Tucker, and John Grim have been very active in providing a niche for this new approach among a vast array of academic environment (Nasr, 1993; Tucker & Grim, 2001). Ever since, we assume that ecotheology has proved itself as a plausible reaction against the ecological crisis that our world has been going through. In this context, different calls were issued by religious leaders from different spectrums calling their disciples to be more environmentally friendly.

However, literature, and especially the one with a religious background, was not amply perceived as a possible vessel for an ecological discourse. It is high time literary scholars should give more room to literature in their writings and believe in its ability to provide a helping hand in curbing the ecological crisis. This issue is not given the space it deserves be it worldwide or nationwide. From a Moroccan perspective, several writers have tackled environmental issues including the late Farid Alansari, a contemporary religious scholar of Arabic expression who passed away in 2009. His writings and thought were put under meticulous scrutiny from religious, literary and political backgrounds. Still, very few scholars or researchers to the best of my knowledge have targeted ecotheological aspects in his writings. It is true that al-Ansari's novels are not ecological in the first place; yet, perusing them has enabled us to unearth the environmental side of which they abound.

Thus, the purpose of this chapter is to track the ecological discourse in Farid Alansari's The Return of the Cavaliers from an ecotheological perspective. Several readings of the aforementioned novel allowed us to come across a variety of ecological themes. These themes oscillated between non-human animals, food and vegetarianism, ecofeminism, overconsumption, secularism, ecotheology, ecotherapy, in addition to communion with nature. Nonetheless, since we were invited to provide only a chapter in this academic book on the environment, our contribution will be confined to two major axes—namely communion with nature and ecotherapy. These two themes will be analyzed from an ecotheological perspective to find out the nature of al-Ansari's discourse towards environment taking into account his cultural and religious backgrounds.

Method

This chapter analyzes the literature written by the Moroccan scholar and novelist, Farid Alansari. To accomplish this task, *The Return of the Cavaliers* is analyzed and deeply scrutinized in order to track any likely ecological discourse. Bearing this in mind, ecological traces in this novel are examined and excavated. Choosing a specific method to analyze a literary work cannot always be justified or defended. Apparently, methodologies, despite being varied and different, have their own points of strength and weakness (Dawson, 2009). In this respect, David Silverman argues that there are no right or wrong methods, but appropriate ones for one's research topic (Silverman, 2013). That said, among the vast panoply of approaches that a researcher can rely on in their data analysis, we assume that both a qualitative approach and an ecotheological approach will be reliable means to analyze and sift through the numerous data that have been collected throughout the realization of this chapter.

Sampling

Sampling data is considered one of the most important steps in conceptualizing a sample design. According to Maxwell, it "[r]efers to the decisions you make in relation to from whom you will gather the data you need to answer your research questions" (Ravitch & Mittenfelner, 2016, p. 137). Taking this into account, we decided to look for literary works that not only answer research questions, but also bridge the gap that the chapter intends to fill in. In this context, we did explore through Moroccan literature with an Islamic background and opted for Farid Alansari's literary works as a sample for our case study. Among his literary heritage, we decided to opt for *The Return of the Cavaliers* as a case study. The criteria for choosing this novel are twofold: Firstly, the novel under probe is Al-Ansari's only novel that was translated into English, a reason that saves us both time and effort for translation. Secondly, thoroughgoing readings of The Return of the Cavaliers enabled us to find out the abundance of themes related to environmentalism and ecotheology as it will be pointed out in the finding section.

Data Collection and Analysis

The data that served as themes of discussion and analysis were collected principally from al-Ansari's novel under scrutiny and underwent a thematic analysis. "[W]hen data is analyzed by theme; it is called thematic analysis. This type of analysis is highly inductive, that is, the themes emerge from the data and are not imposed upon it by the researcher" (Dawson, 2009, p. 119), a process we have opted for throughout our data collection. This process assisted us in unearthing different themes serving the purpose of our study. The themes that came to the fore emanated from the data collected and not the other way around. Gibson and Brown corroborate this concept when they state that "[t]hematic analysis involves noting relationships, similarities, and differences in the data. A theme reflects important concepts in the data and is often understood as 'a generalized feature of a data set' "(Ravitch and Mittenfelner, 2016, p. 219; emphasis is original). Auberbach and Silverstein likened this process to moving through a staircase that starts with "[r]aw text, then identifying relevant texts, then finding repeating ideas, and then grouping these into themes" (Auberbach & Silverstein, 2003; Riger and Sigurvinsdottir, 2016, p. 34).

The themes extracted from our data underwent a process of coding. This process was carried out manually, a process that included underlining, circling, and coloring key words and expressions. We also wrote pertinent hypotheses, questions, and our first impressions in the margins, and marked contradictory points. The emerging themes were organized in charts, each theme next to the page where it was mentioned to facilitate the issue of reference. Thus, despite being very simple and not requiring too much effort, the coding stage has been very useful in our data collection phase since it was both time and effort-saving.

Issues of Validity

"[A]ll research is concerned with producing valid and reliable knowledge in an ethical manner" (Merriam & Tisdell, 2016, p. 237). In this chapter, we tried to reach the same objectives although "[v]alidity in qualitative research can never be fully ensured" (Ravitch & Mittenfelner, 2016, p. 189). However, Ravitch and Mittenfelner (2016) believe in the researchers' ability to increase the scope of validity in their research (p. 189). In the same vein, Merriam and Tisdell (2016) assert that "[t]he validity and reliability of a study depend upon the ethics of the investigator" (p. 260).

Validity is also referred to as trustworthiness, quality, or rigor. These concepts are sometimes used interchangeably (Ravitch & Mittenfelner, 2016, p. 188). In this regard, and as a means to boost these characteristics in this chapter, we managed to apply the same methodology applied to two other novels penned by Farid Alansari, namely The Last of the Cavaliers and Unveiling the Hidden. Consequently, a thematic analysis of the aforementioned novels enabled us to find out that nearly the same themes that were collected in our data happened to pop up in the three works. That said, triangulation was an efficient method to amplify the validity of our study.

Results

Thorough readings of *The Return of the Cavaliers* demonstrated several findings concerning ecotherapy and communion with nature. Our results cast a new light on these issues. By way of illustration, different passages in this regard were incorporated in the novel under probe. The data collected and analyzed in the previous section have enabled us to come up with noteworthy findings, many of which are closely pertinent to our research questions and hypotheses. In this section, we will objectively present and describe our findings and say to what extent they will support our research aims. Therefore, all the data collected "[w] ill be presented in descriptive, narrative form rather than a scientific report" (Creswell & Creswell, 2018, p. 291).

Mapping Ecotherapy in *The Return of the Cavaliers*

If any ecological theme has generated so many controversies as to whether it is an instrumental exploitation of nature or it is an environment-friendly behavior, it must be ecotherapy. One may wonder whether this approach vis-à-vis nature is utilitarian or not as it overlaps with the Romantics' view towards nature. In the eighteenth century, poets used to flee what they deemed corrupt societies to seek a safe haven in nature and heal their inner selves. Ecotherapy as a concept is a portmanteau word made of ecology and therapy. Put differently, it is the use of nature for therapeutic reasons. According to Buzzell and Chalquist, "[E]cotherapy is an umbrella term for treatment modalities that include the natural world in relationships of mutual healing and growth, and as such is a form of applied ecopsychology" (Buzzell and Chalquist, as cited in Chalquist, 2009, p. 64).

Ecotherapy and ecopsychology may be used interchangeably because "[i]n recent years, the meaning of the term 'ecopsychology' has evolved to include nature therapy, which asks not only what we do to the earth, but what the earth does for us—for our health" (Louv, 2008, p. 42). From the short review above, key findings with regard to the issue of ecotherapy were elicited. The excerpt below demonstrates how environment can have magical healing power on humans:

> Such were the passing visions of light (*tajalliyat an-Nur*) whose creeks are flowing now between your hands, oh wakeful one! Thus, carry your sticks on your shoulder and depart towards the soul's enlightenment seeking its original springs so that you may enter the time of the conquest and discover the secret of Fethullah's weeping and you will be healed. (Al-Ansari, 2015, p. 11)

Another excerpt from the analyzed novel reveals how water can provide ecothearpeutic effects to al-Ansari's protagonist.

Your remedy, oh sick companion, is in finding the pearl of your secret?
The pearl of my secret! What is the pearl of my secret? How could I know?
He said, "There are pearls within emerald shells! They grow within the depth of the lake of the secrets? (Al-Ansari, 2015/, p. 13)

The third excerpt, which incorporated ecotherapy as combining both traditional medicine and theological treatment, is as follows:

> My sorrow spoke to me revealing that under Cordoba's Mosque there is a hidden pharmacy kept in a concealed box! My remedy is one of its ancient scrolls which is still buried underneath one of the pillars of the farthest mosque (*al- Masjid al-Aqsa*). (al-Ansari, 2015, p. 207)

Sifting through Communion with Nature in *The Return of the Cavaliers*

This concept, which can also be referred to as unison with nature, is vivid proof of how humans can entertain a close relationship with all the components of the natural world, whether they are human or nonhuman. Thus, bonding with the nonhuman world is palpable proof of any person's ecological concerns. In this regard, al-Ansari brought this notion to the fore, proving that the protagonist of his novel, who was born in a rural area, has reached high realms in his relationship with the natural world. The following excerpts show Fethullah's communion with nature. As an illustration, the two excerpts below indicate a reciprocal communion between Fethullah and components of the environment:

> "And for his weeping all the birds in the world would also weep!" (al-Ansari, 2015, p. 3)

> "Birds and sparrows followed his prayers and the chanting of the remembrance of God. The seagulls used to carry the echo of his weeping throughout Anatolia. . ." (Al-Ansari, 2015, p. 19)

Fethullah's communion with nature did not include only birds; it also encompassed flora as shown in the excerpts below:

> He remembers how he was so happy when his father instructed him to go to Korucuk to bring some branches of the willow tree from the family's garden so that they could plant them in front of their new house to remind them of their house in Korucuk. (al-Ansari, 2015, p. 30)

Inanimate objects also have a room in our protagonist's connection with his surrounding environment. The passage below displays Fethullah's love to the mosque where he used to be an Imam:

> Once of the windows seemed to welcome him so he threw himself on it with a wounded heart that he was tired of walking in rotten places. When he entered through the window and closed its door, he felt its wall kindly and warmly embracing him as if it were a loving and caring mother. (al-Ansari, 2015, p. 127)

While each of the aforementioned excerpts mentioned each component of Fethullah's environment separately, the excerpt below seems to enclose all of them en masse:

> Whenever he would sit underneath the domes and minarets, flocks of doves and hoopoes would land listening to him chanting the spring's hymns. When they comprehended the lesson, they would then return to him holding branches of olive trees between their beaks and some grains as evidence as had arrived at the lad of peace! (Al-Ansari, 2015, p. 224)

Discussion

The data collected from the analyzed novel enabled us to find out al-Ansari's faith in nature's healing power, especially water and religious sites. He introduced them as having effective therapeutic impacts on the narrator's ailing body. Here again, the environmental and the theological intertwine to provide an ecotheological view of the concept of therapy. Additionally, al-Ansari provided some tacit hints in this concern like "tajalliyat an-Nur," soul's enlightenment, "al-Masjid al-Aqsa," and the Mosque of Cordoba, referring us and the readership to his Sufi and religious backgrounds. However, al-Ansari did not mention nor did he specify how this ecotherapeutic process can take place. Thus, it is up to nature to apply its therapeutic effects to the novel protagonist without any human intervention. With regard to water, this ecotherapeutic source was revealed to our protagonist by a person called "the master." The term and the context in which it was brought may imply a religious significance. Here again, al-Ansari combined theology with ecology since he chose a person of a religious background to carry out the healing process. Therefore, incorporating "the master" in healing humans entails that the environment, when put between "pious" hands, can accelerate this healing procedure.

However, despite being a major theme in the field of environmentalism, al-Ansari's use of ecotherapy was very limited. Firstly, this use was a passive form of therapy since it is confined to "connecting with the aesthetic beauty of natural environments that in themselves become places of healing" in contrast to active ecotherapy, which is "a more active engagement in which therapy is conducted using the resources of natural environments" (Kamitsis & Simmonds, 2017, p. 2). Secondly, al-Ansari was very restrictive in linking the concept of ecotherapy only to water and mosques. Other scholars stated that ecotherapy can take place via "green infrastructure and exercise, nature in healthcare settings, animal-assisted therapy, horticultural therapy, and outdoors restoration" (Chalquist, 2009, p. 64). Finally, the process of ecotherapy is presented to take place only outdoors. Thus, one may wonder if nature and its healing power can be incorporated in the counseling room. In this context, Kamitsis and Simmonds (2017) opined that ecotherapy can take place in indoor settings as well.

With regard to communion with nature, the data collected from the analyzed novel empowered us also to find out how it can be a great step forward in shaping our environmental behavior. The environment will no longer be regarded as a fetishized other; rather, it will be perceived as a neighbor or a kin whom should be treated with all the respect it deserves. In this respect, Fethullah proved to entertain an environmentally friendly relationship with nature both during his boyhood and as an adult. This unconditional endearment did not distinguish between the animate or the inanimate world. While young Fethullah's environmentalism can be interpreted as a usual childlike love for his surroundings, his concern and care as an adult proved the protagonist's ecological concerns.

The way al-Ansari dealt with communion with nature converges with several scholars in the field of ecotheology. In this context, Tucker (2003) Tucker urges us to "celebrate our kinship not only with other humans but also with all life forms" (p. 11). Such empathy will prove that "we are not simply human beings but universal beings" (2003, p. 11). On his part, Roger S. Gottlieb promotes the issue of neighborhood departing from the Biblical verse "Love thy neighbour." He contends that if we really love our neighbours, we should not put their lives in jeopardy even if they live miles away (2006, p. 45). Our neighbour is no longer the person who lives next door; rather, "our neighborhood is the entire earth" (2019, p. 26). If the notion of neighborhood can include the non-human world, then "Thou shall not kill" can be applicable in this context as well. Gottlieb (2019) asserts that orders of the type "Do not murder your neighbor" should be used in the environmental context.

Andy Bell believes in a broader notion of neighborhood as well. For him, it should include "the four corners of the world" (2013, p. 387). Humans should reconsider their uneconomical misconduct even if they think it is not nefarious to the environment. Bell (2013) suggests that "when I flip on a light switch that causes coal or oil to be burned, I am responsible for how the toxins from these smoke stacks harm my neighbors" (p. 387).

This ecological consciousness is not a novelty; it was prevalent in Romanticism. Bron Taylor mentions that Muir, for instance, "regularly expressed a spirituality of belonging and kinship" (2010, p. 68). Tony Walting expresses a nexus between being a neighbor of creation and being a steward. According to him, a human who is really a neighbor of nature will exercise "power over nature with humility and praise" (2009, p. 146). Kinship with nature can thus reach its zenith to become an indivisible bond with it. Gottlieb thinks that such a communion can be possible. He came to the conclusion that sifting "through Ecotheology is the sense that we live communally with nature that we can communicate with" (2006, p. 43). On his part, Peter Singer states that Saint Francis De Assisi was the concrete epitome for nature communion. He addressed all creatures as siblings. For the latter, "The sun, the moon, wind, fire, all were brothers and sisters to him" (1975, p. 288). Singer seems to share this Franciscan stance when he refers to non-human beings, for instance, as "our closest nonhuman relatives" (1975, p. 18).

Such an approach, namely eco-theology, designs religion as a possible carrier of ecocentrism. Tucker (2003) maintains that humans can shift away from anthropocentrism to "ecocentric or even anthropocosmic ethics" (p. 48). Like eco-theology, dark green ecology proponents underscore its ecocentric attitudes. Taylor (2010) suggests that it "is generally deep ecological, biocentric, or ecocentric, considering all species to be intrinsically valuable, which is valuable from the usefulness to human beings" (p. 13). The religious and the environmental sometimes intersect in Gottlieb's writings. He found out that espousing deep ecology will allow us to "see nature as a mother, a lover or a partner" (2006, p. 514).

Limitations of the Study

The current study has some limitations, as with the majority of studies. The first limitation concerns the scarcity of English-written articles about Alansari's literary work from ecological or ecotheological perspectives. Moreover, drawing valid and reliable conclusions from one writer with a religious background cannot help one generalize the findings on all the literature with a religious background. Additionally, our analysis was conducted on the English translation of the novel and not the original script. Going back to the original version of the novel enabled us to find out some dissimilarity between the two versions. By way of illustration, Amany Shalaby provided "a place of washing for the sick and the low-spirited" as a translation for a phrase where Prophet Job was mentioned in the Arabic version. Thus, the absence of the name "Job" in the English translation hampered us from citing this excerpt; its presence in the English translation would have otherwise allowed us to link it to ecotherapy from a more comprehensive ecotheological perspective. These limitations are not to be seen as points of weakness for our study; rather, they can be perceived as a call for scholars interested in the field of Ecotheology to conduct in-depth inquiries on this issue.

Conclusion

By limiting the process of environmental healing effects mainly to water and mosques—key elements from an Islamic perspective—al-Ansari highlighted the moralistic value of nature in addition to its aesthetic one. Nevertheless, we could not trace any sorts of negativistic or dominionistic[11] values in his ecological discourse. Yet, nature is still perceived from its utilitarian benefits to humans proving some degree of instrumentalism in al-Ansari's discourse. It is true that al-Ansari's religious heritage made his protagonist treat environment with awe and respect through communion with nature; still, the issue of ecotherapy makes us doubt his ecocentric stances and environmental attitudes which are yet to be fully adopted by ecotheologists.

References

Al Ansari, F. (2015). *The return of the Cavaliers: Biography of Fethullah Gulen* (A. Shalaby, Trans.). Blue Dome Press.

Auberbach, C. F., & Silverstein, L. B. (2003). *Qualitative data: An introduction to coding and analysis*. New York University Press. https://doi.org/10.18574/nyu/9780814707807.001.0001

Bell, A. (2013). An eco-theology. *Crosscurrent*, *63*(4), 386–389.

Chalquist, G. (2009). A look at the ecotherapy research evidence. *Ecopsychology, 1*(1), 64–75. https://doi.org/10.1089/eco.2009.0003.
Creswell, W. J., & Creswell, J, D. (2018). *Research designs: Qualitative, quantitative and mixed methods approaches.* Sage Publications, Inc
Dawson, C. (2009). *Introduction to research methods: A practical guide for anyone undertaking a research project).* How To Content
Gottlieb, R. S. (2006). *Religion, nature, environmen*t. In A greener faith: Religious environmentalism and our planet's future (pp. 19–56). Oxford University Press. https://doi.org/10.1093/acprof:oso/9780195176483.003.0002
Gottlieb, R. S. (2019). *Environmental crisis and moral life. Morality and environmental crisis.* University of Birmingham.
Kamitsis, I., & Simmonds, J. G. (2017). Using resources of nature in the counselling room: Qualitative research into ecotherapy practice. *International Journal for the Advancement of Counselling Article, 39*, 229–248. https://doi.org/10.1007/s10447-017-9294-y
Louv, R. (2008). *Last child in the woods.* Algonkin Books.
Merriam, B. S., & Tisdell, E, J. (2016). *Qualitative research: A guide to design and implementation.* Jossey- Bass
Nasr, S. H. (1993). *The need for a sacred science.* Richmond: Curson Press
Ravitch, S. M., & Mittenfelner, N. (2016). *Qualitative research: Bridging the conceptual, theoritical and methodological.* Sage.
Riger, S., & Rannveig Sigurvinsdottir, R. (2016). Thematic analysis. In E. L. Jason & D. S. Glenwick (Eds.), *Handbook of methodological approaches to community-based research.* Oxford University Press.
Silverman, D. (2013). *Doing qualitative research.* Sage.
Singer, P. (1975). *Animal Liberation.* Open Road.
Taylor, B. (2010). *Dark Green Religion.* University of California Press.
Toynbee, A. (1972). The religious background of the present environmental crisis. *International Journal of Environmental Studies*, 3(1–4), 141–146. https://doi.org/10.1080/00207237208709505
Tucker, M. E., & Grim, J. (2001). Introduction: The emerging alliance of world religions and ecology. *Daedalus*, 130 (4), 1–22.
Tucker, M. E. (2003). *Worldly wonder: Religions enter their ecological phase.* Open Court.
Watling, T. (2009). *Ecological imaginations in the world religions: An ethnographic analysis.* London: Continuum.
White, L. (1967). The historical roots of our ecologic crisis. *Science*, 155(3767), 1203–1207. https://doi.org/10.1126/science.155.3767.1203

Conclusion

Mohamed Mliless

As a general conclusion of this book, *Interdisciplinary Approaches to Environmental Identity and Portrayal* synthesizes the diverse themes explored throughout the work, emphasizing the complex interplay between humans and their environment. This scholarly work crosses multiple disciplines, providing a comprehensive understanding of how our perceptions and portrayals of the environment are shaped. It highlights the critical narratives surrounding food production and animal welfare, illustrating their profound implications for our environmental relationships. The unconventional role of the military in addressing climate change is presented as a unique perspective on environmental stewardship, challenging traditional notions of responsibility.

The linguistic exploration within the book reveals how language not only reflects but also influences our understanding of landscapes, positing them as sentient entities deserving of recognition. The intersection of environmental protection and technology underscores the transformative potential of digital tools in conservation efforts.

Legal frameworks governing environmental protection are meticulously examined, with a case study on Morocco showcasing the balance between the right to a healthy environment and the complexities of international and national laws. Historical analyses, particularly of forest policy in India during colonial and post-colonial periods, provide context for contemporary environmental challenges.

The psychological and societal dimensions of environmental interaction are explored, revealing how our identities and communities are shaped by our environments. The book also addresses pressing issues such as the environmental crisis, postcolonialism, and gender inequality, particularly in the Global South, while emphasizing the vital role of women in conservation efforts.

Through critical analyses of literature and the therapeutic benefits of nature, the book advocates for a deeper communion with the environment, highlighting its significance for mental health and well-being. Each chapter contributes a unique perspective, making this work an essential resource for those seeking to understand the multifaceted relationship between humans and the environment from various disciplinary angles.

Interdisciplinary Approaches to Environmental Identity and Portrayal serves as a vital exploration of the intricate connections that shape our environmental identities, urging readers to consider the broader implications of their interactions with the world around them. The chapters of *Interdisciplinary Approaches to Environmental Identity and Portrayal*" intersect in several significant ways, emphasizing the multifaceted nature of environmental issues and the interconnectedness of various disciplines. For instance, the chapters "Global Climate Change: From Anthropocene to Capitalocene" and "Emotions and Disease: The Negative Framing of Non-Human Animals in Climate Change Discourses" both address the narratives surrounding climate change. The former critiques the Anthropocene framework, suggesting that capitalism is a primary driver of ecological crises, while the latter examines how these narratives negatively impact the perception of nonhuman animals in climate discussions. This connection highlights the need to reconsider how we frame environmental issues within socio-economic contexts, urging a shift in perspective that considers the implications of our narratives.

Transitioning from these overarching themes, the book further explores how language and legal frameworks shape our understanding of environmental identity. Chapters like "Grammatical Reflections of Sentient Landscapes: Linguistic Evidence for Environmental Personhood" and "The Law of Hazardous Waste and Sustainable Development" intersect by emphasizing the necessity of recognizing landscapes as sentient. The former provides linguistic evidence for this recognition, while the latter discusses the legal obligations that arise from such an understanding, particularly in the context of sustainable practices. This interplay between language and law underscores the importance of framing environmental issues in a way that acknowledges the rights of both human and non-human entities.

Moreover, the book delves into critical issues of water rights and human security, as illustrated in "The Protection of Water Rights in the Face of Human Security: the Case of Morocco" and "The Right to a Healthy Environment in Morocco between International Conventions and National Legislation." Both chapters focus on Morocco, highlighting the tension between local rights and international legal frameworks. Here, the importance of water rights emerges as a critical aspect of both human security and environmental health, emphasizing the need for policies that prioritize access to clean water as a fundamental human right.

In addition to these legal and linguistic dimensions, the book provides historical context that informs modern environmental challenges. For example, "Forest Policy and Management in Colonial and Post-Colonial India: A Historical Study" and "The Environmental Crisis, Postcolonialism, and Gender Inequality in the Global South: A Postcolonial Ecofeminism Approach" connect by examining how historical policies have lasting impacts on contemporary environmental issues, particularly concerning gender and postcolonial identities. This historical lens enriches our understanding of current struggles, illustrating how past injustices continue to shape present realities.

The exploration of visual and emotional representations further enhances the book's depth. Chapters like "Visual Language and Environmental Representation in Moroccan Cinema *Sotto Voce* as a Case Study" and "The Representation of Hunted Animals on Social Networks: A Visual and Critical Discourse Analysis" investigate how visual media shapes public perceptions of environmental issues. They highlight the emotional responses elicited by these representations, which can significantly influence activism and policy decisions. This focus on visual culture underscores the power of imagery in shaping our understanding of environmental identities.

Additionally, the psychological aspects of environmental interaction are examined in chapters such as "The Rain Sound as an Affective Force that Unites Humans and Nature" and "Ecotherapy and Communion with Nature in Farid Alansari's '*The Return of the Cavaliers*': An Ecotheological Perspective." These chapters delve into the therapeutic benefits of nature, suggesting that emotional and psychological connections to the environment can foster a sense of identity and community. This exploration emphasizes the importance of nurturing these connections as a means of promoting mental health and environmental stewardship.

In light of these intersections, *Interdisciplinary Approaches to Environmental Identity and Portrayal* is crucial for a diverse audience. It serves as a comprehensive resource for academics and researchers interested in environmental studies, sociology, law, and psychology, providing interdisciplinary insights that enhance understanding of complex environmental issues. Furthermore, the book equips policymakers and environmental activists with the knowledge needed to advocate for effective environmental protection strategies, addressing legal frameworks, historical contexts, and contemporary challenges.

Students and educators will also find the diverse topics covered in the book to be an excellent educational resource, fostering critical thinking about the relationship between humans and their environment. Additionally, the book raises awareness among the general public about the interconnectedness of environmental issues, encouraging readers to reflect on their roles and responsibilities in shaping a sustainable future.

Index

List of Editors and Contributors

Abdelaziz Dqaqi, a middle school teacher for twenty years, holds an MA in Green Cultural Studies. He is a fourth-year doctoral student in the Space and Culture Research Team at Mohamed the First University. His academic interests include Ecotheology and Postcolonial Studies. His doctoral thesis focuses on religious environmentalism in Farid Alansari's novels from an ecotheological perspective. He has actively participated in conferences and study days, delivering presentations on Environmental Theology and Postcolonial Discourse.

Chougrani Elhoucine is an international law professor at Cadi Ayyad University, Morocco. He is the Director of MOROFUGE and a member of the International Research Center for Crisis Management. His research focuses on international environmental law, with recent publications on water, climate justice, and the rights of future generations. He was born in Morocco in 1971 and holds a PhD on the European Investment Bank's role in environmental protection. Additionally, he has experience as an Accountant General at the Ministry of Economic and Finance from 2002 to 2014.

David Stringer is an Associate Professor of Second Language Studies and Director of Undergraduate Studies, Second Language Studies Indiana University. His main research area is the acquisition of lexical semantics and syntax, with a focus on universals of language and cognition. Additional areas of interest include language attrition, multilingualism in postcolonial societies, and biocultural diversity conservation, linking indigenous language revitalization to the conservation of ecosystems. His interests extend to all formal aspects of second language acquisition and multilingualism, and he enjoys working with students on a wide range of projects.

Diego Luis Forte is a professor in Linguistics and Social Semiotics and a researcher in the National Library of Argentina. He has taught courses in the University of Buenos Aires, the National Institute of Public Administration of

Argentina and teaches postgraduate courses on multimodality in the University of La Serena, Chile. His main lines of work involve the place of faces and music in Systemic-Functional Multimodality, Social Semiotics, Ecolinguistics and the discursive construction of speciesism in Argentina.

Douae Serraf is a PhD researcher at Mohamed I University in Oujda, Morocco studying intersectional literary studies, gender studies, women and fiction, and environmental literary studies. She is also a part-time EFL instructor in Oujda and holds a MA degree in Women and Gender Studies. Douae is a PhD candidate in the Communication, Education, Digital Usage, and Creativity research laboratory at Mohamed I University and represents UNICEF at MUN conferences.

El Madani El Mehdi holds a PhD in Applied Linguistics. His doctoral thesis is entitled "The Impact of Metacognition on Language Learning." He received his MA in Applied language Studies from Moulay Ismail University, Meknes Morocco. He works as a part-time teacher at the department of English studies at the school of Arts and Humanities, Meknes. His major research fields of interests are Applied linguistics, Foreign Language Learning and Teaching, and Metacognition.

Fatima Zahra Alaoui Mahrez is a Doctoral Student in the Space and Culture Research Group-CEDUC Laboratory at Faculty of Letters and Human Sciences-Mohammed I University, Oujda-Morocco. She is specialized in Intercultural Communication and International Relations. She is a temporary professor at Mohammed I University. She has participated in many scientific meetings related to environmental issues. Her major areas of interest are communication, international relations, public policies, and critical discourse analysis.

Fouad Akki is an Assistant Professor of Applied Linguistics at Moulay Ismail University in Meknes, Morocco. He holds a PhD in Applied Linguistics and his research interests include language skills, curriculum development, discourse analysis, ecolinguistics, bilingualism, and multilingualism. Dr. Akki has authored and co-authored articles on teaching speaking and writing skills, English for Specific Purposes, Ecolinguistics, and Arabic-English-Arabic Translation. He has reviewed articles for many international journals.

Fouad Souiba earned a degree in communication and journalism after studying in Morocco and Quebec. He has been a journalist in the Moroccan and international press since 1982. He is a radio and television columnist who has presented and produced well-known programs. He has contributed to the establishment of various events and the promotion of cinema and television as a film festival presenter. He is the editor-in-chief of several journals, including *Cinémasrah* and *Al Mashhad*. He has also directed eleven documentaries for Al Jazeera and other features for the big screen.

Hafsa El Bastami holds a BA in International Studies and an MA in International Studies and Diplomacy from Al Akhawayn University. She is pursuing a doctoral degree at Ibn Tofail University, focusing on women's empowerment and gender issues. She has written extensively on women's organizations and empowerment. Hafsa's professional experience involves managing exchange and research programs in higher education and international cooperation between Morocco, the US, and European countries.

Hajar Taha is a dedicated researcher specializing in environmental humanities. She holds a PhD in English Studies from the University of Mohammed First in Oujda, Morocco. Dr. Taha has contributed to the field through several publications on ecofeminism and environmental issues. She is also an active member of the Association of Cultural Ecology and Communication (ACEC) in Morocco.

Hicham Laabidi is an associate professor in the Department of English Studies at the School of Arts and Humanities, Moulay Ismail University, Meknes, Morocco. He worked as a teacher-trainer in Teachers' Training Centre (CRMEF), Meknes before joining the university. He participated in different national and international conferences. He published several articles in various international journals. Textbook evaluation, ICT integration in education, and professional development are his major research areas.

Hind Hassine is a Doctoral student in the Civil and professional systems Laboratory at Faculty of Law, Economy and Social Sciences- Mohammed I University, Oujda-Morocco. She is specialized in privet law and interested in topics related to environmental protection and sustainable development.

Hind Majdoubi is a tenured professor of law at Ibn Tofail University, Kenitra, Morocco, and an associate professor at the Faculty of Legal, Economic and Social Sciences of Souissi Rabat. Dr. Majdoubi works mainly in the field of environmental protection and sustainable development. Her research has focused on environmental protection, the fight against waste and action against vulnerabilities and discrimination among other issues. In this context, she has made several communications and publications.

Jaouad Achtitah (PhD, Mohammed I University, Morocco) is a professor of English at Ibn Tofail University, and Chair of the Association of Cultural Ecology and Communication in Morocco. His research interests revolve around ecological psychology, affect studies, and environmental criticism and literature. He participated in many national and international seminars. He edited the book of Proceedings of the 1st National Environmental Humanities Conference: Perspectives from Morocco, and his research appears in book chapters including Current Issues in Environmental Discourse, Eco-Anxiety and Pandemic Distress: Psychological Perspectives on Resilience and Interconnectedness, and Ecofeminism and Allied Issues.

Khadija Anouar is a tenured professor of law at Sidi Mohamed Ben Abdellah University Fes, Morocco, and an associate professor at the faculty of legal, Economic and Social Sciences of Souissi Rabat. Dr. Anouar works mainly in the field of environmental protection and sustainable development. Her research interest is devoted to issues of renewable energy, environmental protection, waste, labor law, criminal law. She teaches labor law, waste law, and criminal law. She has authored and co-authored invaluable publications and communication in the aforementioned disciplines.

Latifa Belfakir, now the chief of the English department at the faculty of humanities in FLDM, has had a long and fruitful teaching experience in Meknes before she started her journey at the faculty of Letters and Human Sciences in Fes. Dr. Belfakir is presently both enjoying the education profession as a full professor in the department of English studies and as the coordinator and leader of the Master program: Language, Communication & Society. She is (co-)author of a series of books, book chapter and indexed article in international journals. She is overall keen on enhancing her students' soft skills and encourages extracurricular activities on a par with the core or content courses.

Mohammed Adli is an Associate professor at High Institute of Technology in Oujda, Morocco. He got his Doctorate degree in English studies on May 2016. Professor ADLI is very interested in the fields of Social Sciences and Humanities. He has also participated in various study days and seminars in Morocco and overseas and published various publications in his area of study.

Mohammed Larouz is the Dean of the School of Arts and Humanities of Meknes and Vice-President of Academic Affairs at Moulay Ismail University of Meknes. He earned a PhD in Applied Linguistics from Fez University in 2004 and has worked as a Professor in the Department of English Studies since 2005. Dr. Larouz also holds an MA in Applied Linguistics from Essex University and has been a Fulbright Scholar to the USA twice. He has research interests in TEFL, sociolinguistics, research methodology, and communication.

Mohamed Mliless is an independent researcher in Ecolinguistics and Critical Environmental Discourse Analysis. He holds a PhD in Applied Linguistics from the University of Moulay Ismail of Meknes, Morocco. He has published books on legal and environmental discourse and authored articles on social and environmental issues. Dr. Mliless is a member of the International Ecolinguistics Association (IEA) and serves on the boards of several journals.

Dr. Mohammed Moubtassime holds a PhD in Comparative Linguistics from Sidi Mohamed Ben Abdellah University, where he teaches at the Faculty of Humanities. He holds leadership positions and is a member of the national board for evaluation and accreditation at the Moroccan Ministry of Higher Education. Dr. Moubtassime has expertise in educational management and curriculum development. His research interests include linguistics, curriculum development, teaching and learning, ICT, educational management, and more. He has organized conferences, published books and articles, and serves as the Director of the Doctoral School at Sidi Mohamed Ben Abdellah University.

Nasreddine EL Guezar holds a PhD from the Faculty of Letters and Humanities, Mohamed I University, and a part-time EFL instructor at the American Language Center in Oujda, Morocco. Nasreddine obtained his MA in Green Cultural Studies, and he is currently in the process of writing his doctoral thesis on Science and the Humanities in Michael Crichton's Science Fiction Novels, Sphere, Prey, and Next under the supervision of Professor Said Mentak. He is expected to finish writing his doctoral thesis by the end of 2023. Nasreddine contributed to numerous journal articles and academic national and international conferences. At Mohammed I University, he also taught academic subjects and received extensive pedagogical training. He is interested in researching the relationship between science and the humanities, ecological literary criticism, Anthropocene fiction, speculative fiction, environmental awareness in education and in EFL classrooms, and ecocinema.

Prashant Tripathi was born in Siddharth Nagar a small district of India on August 19, 1997. He completed his graduation from Hindu College, University of Delhi in BA History (Hons.). Then he moved to Banaras Hindu University in 2016 and gets my MA in History. Currently, Prashant Tripathi is a PhD student at the Department of History, Banaras Hindu University. He has published many articles, papers in UGC approved journals & peer-reviewed journals and has presented many papers in seminars, conferences organized by prestigious universities and organizations.

Rabiaa Bourhim holds a PhD in Public Law and Political Sciences. He is a member of the International Center for strategic research on territorial governance and the Research Center for Environment, Human Security & Governance (CERES). With expertise in government procurement, his research focuses on environmental change, conflict management, sustainability, and public policy. Dr. Bourhim has published numerous scientific articles, participated in conferences, and serves as a consultant to the parallel government of youth in Morocco.

Redouane Bougtib is a doctoral researcher. Redouane is an assistant at Mohammed V University and an assistant researcher in Ibn Khaldoun project. Graduated from Ibn Tofail University with majors in Cultural Studies, Redouane participated and contributed in the organization of national and international conferences, lectures and workshops. Redouane has also worked as a high school teacher of English for four years.

William R. Naugle is an Assistant Professor in the Department of Languages and the Director, Office of International Relations at Al Hussein Technical University (HTU) in Amman, Jordan. His doctorate is in Applied Linguistics from New York University, where his initial researched focused on hedging within the language of the law. Language acquisition plays a central role in his current research efforts, collaborating with colleagues in Spain and the US on neurodivergent language development and language learning for vocational technical students throughout Jordan. He acquired Italian and German as heritage languages and communicates in French, Arabic, and Spanish. Dr. Naugle is an English Language Specialist for the US Department of State and a Fulbright TEFL Scholar alumnus.